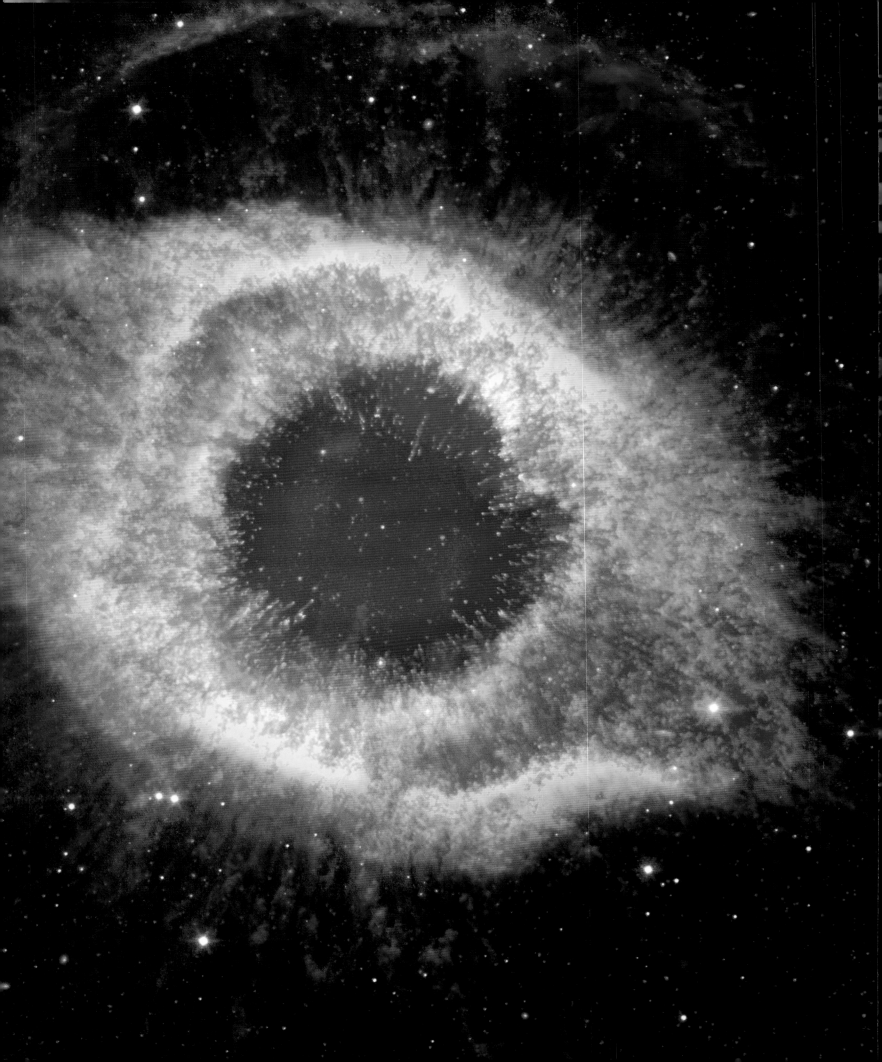

FOREWORD BY
ARTHUR C. CLARKE

THE HISTORY OF
ASTRONOMY

Heather Couper & Nigel Henbest

**To Bernard Henbest
1924-2004
An inspirational mind, a brilliant scientist
and a much loved father**

First published in 2007 by Cassell Illustrated, a division of Octopus Publishing Group Limited,
2-4 Heron Quays, London E14 4JP
An Hachette Livre UK Company

Text copyright © 2007 Nigel Henbest and Heather Couper
Design and layout © Cassell Illustrated

A CIP catalogue record for this book is available from the British Library.

ISBN: 978-1-84403-570-0

Commissioning Editor: Laura Price
Project Editor: Fiona Kellagher
Assistant Editor: Eoghan O'Brien
Design: Richard Scott
Production: Caroline Alberti

Printed in China

CONTENTS

Foreword

For more than half a century, I have been lucky enough to observe – and occasionally experience – the greatest revolution in the history of astronomy, perhaps the oldest of all the sciences. With the dawn of the Space Age in 1957, it moved from observation to experimentation – and now exploration.

But astronomy is not just about space probes, satellites, space telescopes and other expensive toys deployed by rich nations often trying to outdo each other. At the most basic, it's about exploring that great laboratory within easy access to anyone, anywhere on the planet: the night sky.

We have to admit though, that space exploration has expanded our knowledge and understanding of the heavens as never before. We have learnt more about the universe in the past half century than we did for several centuries since Galileo turned a refracting telescope on the night sky circa 1609.

For much of history, astronomy involved peering through ground-based optical instruments – no matter how powerful they were, they had one insurmountable limitation: observations were distorted by the Earth's atmosphere. But as Heather Couper and Nigel Henbest remind us in this well-researched history of astronomy, this did not prevent generations of astronomers from carrying out some of the finest observational work.

When I joined the British Astronomical Association (BAA) in the 1940s, any talk of space travel was regarded as hair-brained nonsense, fit only for boys' magazines and the cheaper science fiction 'pulp' magazines. It is now amusing to recall some of the objections raised against the idea that we would one day be able to leave the Earth. In a famous (or infamous) editorial, The *New York Times* once castigated Robert Goddard for thinking that a rocket could work in a vacuum, when 'there was nothing to push against.' Though they apologised in the special Apollo issue on July 17, 1969, by then Goddard had been dead for years.

I also recall a wonderful headline in a British newspaper during the 1930s 'We are prisoners of fire!' when radio echoes revealed that the temperature in the Ionosphere was some thousands of degrees. The journalist responsible for this headline obviously didn't understand the difference between heat and temperature. One would soon freeze to death in the Ionosphere, if the only warmth came from the rare thousand degree gas molecules. (For a good example of this distinction, consider those delightful Guy Fawkes Night fireworks that you can hold against your hand, even when they are spitting out showers of incandescent sparks.)

Until quite recently – except for those deluded individuals who believe in horoscopes[1] – it was generally considered that celestial bodies had no influence on this planet – except, of course, in such obvious cases as the Sun and the Moon. Then, during the course of a little more than a decade, it was realised that impacts from space have had a profound effect upon life on Earth. We might not be here today if an asteroid or comet had not wiped out the competition some 65 million years ago.

I can still recall arguments at the BAA meetings over the origin of lunar craters, where one astronomer remarked: 'The presence of central peaks completely rules out the meteoric hypothesis.' We cannot altogether blame him because there are obvious examples of volcanic activity on the Moon, e.g. the crater Wargentin, which is full of lava up to the brim. So when we know one process is at work, it seems unnecessary to look for another. And who would ever have imagined that the brief upward splash which occurs when you drop a lump of sugar into a cup of coffee can be reproduced on a million-fold greater scale – in solid rock!

Compared to the history of space exploration, which completes half a century this year, the history of astronomy goes back several millennia. To compile this brief history of our understanding of the heavens, Heather and Nigel sift through historical records and artefacts belonging to many cultures and civilisations. Like the ancient Greeks did, they ask all the right questions, but some of the answers remain, for the moment, speculative.

A good example is my favourite artefact from the ancient world, the Antikythera Mechanism. Discovered in 1900 by Greek sponge divers, this bronze contraption consisted of a box with very complex gear wheels. It remained something of an oddity until British physicist Derek Price carried out a detailed analysis in the 1950s. He showed that the mechanism was a component of an analog astronomical computer that displayed the position of the Sun and Moon on any particular day of the year. That was impressive enough, but most remarkable was its age: it was dated back to the first century BC!

I played a small part in this investigation. In the late 1950s, I put Dr Price in touch with Dennis Flanagan, editor of *Scientific American*, who persuaded him to write the article which first presented this astonishing device to the general public ('An Ancient Greek Computer', *Scientific American*, June 1959)[2]. Over the years, I continually pestered Dr Price to complete his research, which was finally published in 1974 ('Gears from the Greeks', in the Transactions of the American Philosophical Society)[3].

Meanwhile, in 1965, I was in the Greek capital to attend a space congress, and took time off to personally examine the device. It required three visits and a letter from an admiral to access the item, tucked away at the time in a cigar box in the basement of the National Museum of Athens (I believe it now has a more prominent display). But it was worth the hassle: looking at this extraordinary relic was a disturbing experience.

Though it was more than two thousand years old, it represented a level that our technology did not match until the eighteenth century. Unfortunately, this device merely described the planet's apparent movements; it did not help to explain them. With the far simpler tools of inclined planes, swinging pendulums and falling weights, Galileo pointed the way to that understanding – and to the modern world.

If the insights of the Greeks had matched their ingenuity, the Industrial Revolution might have begun a thousand years before

Columbus. By this time we would not merely be pottering around the Moon; we would have reached the nearer stars.

Well, that is one of history's greatest might-have-beens. I have often wondered what other treasures of advanced technology may lie hidden in the sea.

Tighten your seat belts to go on a fascinating journey through space and time with Heather and Nigel as your guides.

Sir Arthur C Clarke
Colombo, Sri Lanka
May 2, 2007

1 That's almost everyone in Sri Lanka! When asked for my own views I usually reply: "I think astrology is utter nonsense – but then I'm a Sagittarius, and we're very skeptical."

2 On the copy he sent to me, Derek Price wrote: "Please find some more." I am afraid that the most advanced underwater artefact I have discovered is an early 19th century soda water bottle.

3 In 2006, a team of British, Greek and American researchers examined the artefact using the latest in high-resolution imaging systems and three-dimensional X-ray tomography, and was able to decipher many inscriptions and reconstruct the gear functions.

Introduction

'If I have seen further than certain other men, it is by standing on the shoulders of giants.' So wrote Isaac Newton to his fellow scientist Robert Hooke on February 5, 1675.

In one stroke, Newton summarised the whole history of astronomy. It is an edifice built on the endeavours of countless men and women through the millennia; a vast pyramid of human achievement that points towards the sky.

The history of astronomy is so much more than the history of a science. It is a reflection of our culture: an insight into the development of humankind's ideas and ideals. Why else would we call the cosmic firmament 'Heaven,' and populate it with deities – like Apollo the Sun god and Diana the Moon goddess, along with Jupiter, Venus and the other planets? Why else would we map our long-cherished legends onto the sky, making them concrete as the constellation patterns? Why else have civilizations believed that the stars dictated their lives?

Our ancestors built monuments that are aligned with the heavens. From Stonehenge to the great Pyramids, from the native North American structures at Chaco Canyon in New Mexico to the mysterious mounds of Bronze Age Britain, it is clear that – in those un-light-polluted days – the sky was as important to humanity as events on the Earth.

With the passing of the centuries, we can only guess at the motivation behind these grandiose schemes: are we looking at cathedrals to the cosmos? What is certain is that almost every culture has a 'Creation myth' – which involves the simultaneous formation of 'heaven and earth.'

We are on firmer ground when we reflect on how our ancestors used the stars – for timekeeping, calendar-making and navigation at sea. Even today, a small cadre of Polynesian sailors are following in the footsteps of their forebears, who, around 2000BC, started to explore the myriad islands of the Pacific Ocean, using guidance from the heavens.

'When beggars die, there are no comets seen; the heavens themselves blaze forth the death of princes.' In Act II of Julius Caesar, Shakespeare encapsulated our other fascintion with the sky: that heavenly events reflected life on Earth. The ancient Chinese believed that the sky was literally the mirror of the Earth, and that an unwelcome comet or exploding star indicated rebellion in the provinces. Astronomy and astrology were intertwined until the seventeenth century – when science kicked in.

The Greeks were the first to look to the sky with a scientific eye. How big was the Earth? How far away was the Sun? Does the Sun travel around the Earth, or the Earth around the Sun? How far does our cosmos extend? But with the demise of Greek civilisation, the rationalist approach to the heavens virtually died. The flame was kept alive for a thousand years in Arab lands.

Then, in the 16th century, came the first great revolution in astronomy. The Polish canon Nicolaus Copernicus realised that it was easier to explain the motions of the heavens if he dethroned the Earth from its central position, and made it orbit the Sun.

The scene was now set for astronomy to change forever. In 1609, Galileo Galilei turned his 'optick tube' – the newly-invented telescope – towards the sky. Galileo made bold of his findings: that the Earth circled the Sun, and that the heavenly bodies were not perfect. The Moon was pocked with craters; and the Sun was spotty.

His forthright rebuttal of church doctrine led to Galileo being place under house arrest. But his legacy in astronomy and mechanics inspired a young Englishman, Isaac Newton, who was born in the year that Galileo died.

With his formidable mathematical brain, Newton worked out why bodies in space moved in the way they do: there was a new force to be reckoned with – gravity. At last, astronomers could calculate what was going on in the Universe, rather than just predict the future on what had happened in the past.

From Newton's time onwards, the pace of astronomy quickened. In the eighteenth century, a musician-turned-astronomer, William Herschel, literally doubled the size of the Solar System by discovering a new planet – Uranus. And he paved the way forward to exploring the wider Universe, with his investigations into the nature of the Milky Way.

Victorian astronomers had the bit firmly between their teeth. With the invention of photography, they could record their observations for perpetuity; with the invention of spectroscopy – which reveals the composition of stars and planets – they could work out the chemistry of the Universe. And with precision telescopes, they were at last able to measure distances to the stars.

As the twentieth century hoved into view, the astronomical community was getting to grips with the structure of the distant Universe. Was our Galaxy all that existed; or was it just one of billions of galaxies? The latter proved to be the case. And then – in

one of the greatest discoveries of the last century – Edwin Hubble found that the entire Universe is expanding. As a result of the Big Bang, a colossal cosmic explosion which took place 13.7 billion years ago, the galaxies are all flying apart from each other.

And, very recently, a new revolution in astronomy has taken place – one as great as the upheaval in the era of Copernicus and Galileo. New technologies mean that astronomers are no longer limited to simply looking at the sky. They can now tune into the cosmos at a whole range of wavelengths – from hugely energetic gamma rays to low-frequency radio waves.

This recent cornucopia of data on the cosmos has told us – in no uncertain terms – that we live in a violent universe. The safe, predictable, stars and planets of our ancestry have been replaced by wild worlds. Black holes, colliding galaxies, wayward planets and exploding stars are all out there on view.

But astronomers have are also accruing evidence that, despite all this disruption, there could be life somewhere else out there…

This book celebrates our changing perspectives on the Universe. Over the millennia, astronomers have established the true nature of our cosmos. At each stage, our planet Earth has seemed smaller and less significant. This changing perception has put astronomers at loggerheads with philosophers and priests alike.

Today, we have reached a humbling perspective on the Cosmos. The Earth is an average planet, circling a middle-aged star in an unremarkable galaxy. But we can be proud of one achievement: that our planet has developed a life-form that can gaze out into the Universe – and question what it all means.

Heather Couper and Nigel Henbest

Living with the sky

'Spread out!' urges our guide, as we pick our way up the scrubby slope towards
a vast jagged boulder that has fallen, long ago, from the sandstone cliffs above.

Each taking a different route, we reconvene under the giant red rock. Now we can see why our guide didn't want our combined feet to make a trail. Etched into the rock, at the brink of visibility, are shallow markings. They have the air of great age. And even the slightest wear from the curious fingers of a multitude of visitors could erase these petroglyphs forever.

Amid a set of figures carrying strange instruments is a symbol that sets the hearts of us two astronomers pounding: a circle with loops writhing from its edge. Surely this has to be a total eclipse of the Sun, with the delicate streamers of its atmosphere reaching out into space?

'There was a total eclipse of the Sun visible from here in 1097,' confirms our guide, GB Cornucopia. 'Though we can't definitely say that's what this petroglyph is showing. But come and take a look at this…'

Another face of the giant stone bears a faintly inscribed spiral. From here, we look up to the northeast, where a prominent triangular rock stands against the skyline. 'You see the Sun rising right behind that stone at the summer solstice,' explains Cornucopia.

BELOW *Imposing Fajada Butte guards the entrance to Chaco Canyon, the ceremonial and astronomical centre for the ancient inhabitants of western North America. At the winter and summer solstices, and also at the equinoxes, sunlight shines between three giant rocks at the summit of Fajada Butte, throwing bright "sun daggers" onto an intricate set of spiral patterns carved on the rock behind.*

And on the far side of the stone from the spiral, a line of hollowed out footprints leads to a split in the great rock, where a set of holes has been deliberately pocked out of the sandstone. From this sheltered side of the boulder, our view lies towards the horizon where the Sun sets at Midwinter.

We've come to Chaco Canyon – near the Four Corners where New Mexico, Arizona, Utah and Colorado all meet – to try to understand the very roots of astronomy. What was in the minds of people as they first looked up at the sky?

Our quest for astronomical origins had started at the Griffith Observatory in Los Angeles – which has an excellent view over a more contemporary 'star' symbol, the famous hillside 'Hollywood' sign. The observatory's director, Ed Krupp, is one of the world's leading experts on ancient astronomy.

'I don't think we can ever know anything in detail about the minds and motivations of the people who built places like Chaco Canyon and Stonehenge,' opines Krupp. 'What is absolutely absurd is to claim that we know for certain what they were thinking of. But the investigation of ancient and prehistoric and traditional astronomy is still a worthy enterprise. It's essentially an examination of fundamental human response to nature.'

That's why GB Cornucopia moved to Chaco Canyon as a Ranger, after seeing it featured on Cosmos, Carl Sagan's hugely popular television series on astronomy. 'One of the Chacoan people's tools was certainly astronomy,' Cornucopia avers, 'and they were interested in astronomy because anyone living in this harsh environment who does not understand their environment will not survive.'

The trained eye can find any number of alignments at Chaco Canyon that apparently relate to the Sun and the sky. More than that, though the canyon itself is

ABOVE *The bright star to the left of the Moon, in this Chaco Canyon rock painting, may represent a supernova seen to explode in AD 1054. Halley's Comet, which appeared 12 years later, could be the inspiration for the swirling pattern below.*

BELOW *This thousand-year-old petroglyph bears an uncanny resemblance to a total eclipse of the Sun.*

now deserted, descendants of its people still live in the region – the Pueblo peoples of New Mexico and the Hopi of Arizona. Their traditions may hold clues to the astronomical traditions behind the silent stones.

'The ceremonial life of many of the Pueblo people today revolve around their observations of the Sun and Moon,' explains Cornucopia. Every Pueblo settlement has its own Sun Chief, who monitors where the Sun rises and sets, so providing a basic calendar for the farmers.

'Many of the Pueblan ceremonial times revolve around the winter solstice,' Cornucopia adds, 'to do the proper ceremonies so that the Sun will come back into the sky and start another yearly cycle.'

For all his skepticism, Ed Krupp agrees that the Pueblo people may be carrying on an ancient tradition. 'You see many aspects of their life as represented in the archaeological record that persist in the traditional culture. And a few centuries really is not so long for everything to change and to evaporate.'

Cornucopia puts it this way. 'We find these traditions alive and well among the Pueblo people of today. We believe some of these traditions started here in Chaco; perhaps they were even old traditions by the time they were building the Great Houses.'

The Great Houses are the most stunning sights in Chaco Canyon. Their ruins are impressive even a thousand years on. Each is built from millions of small fragments of dark hard sandstone, meticulously fitted together to form a veritable maze of small rooms and round sunken 'kivas'.

'Sometimes a single building covers over 4 acres (1.6 ha), four storeys high, six to eight hundred rooms,' elaborates Cornucopia. 'Early researchers counted up the rooms, and said 20,000 people lived in the canyon. No-one thinks so any more – the environment is just too harsh. The most anybody is willing to say now is about 6000 people – maybe even as few as one thousand.'

LEFT *The Griffith Observatory in Los Angeles is a twentieth century monument to the sky. Successive directors have trained astronauts to recognize the stars, introduced laser planetarium shows and pioneered the study of the astronomy of ancient cultures.*

The extra rooms were probably for visitors – thousands of them who congregated for special ceremonial events. Around AD 1000, this broad and relatively shallow canyon was the hub of trade and culture – and astronomy – for the America Southwest.

People traveled here along a network of paved roads: in this part of the world, all roads led to Chaco Canyon. One of them heads in a due north-south line for 35 miles (55 km), even traveling up vertical cliffs – a sure indicator that it was as much ceremonial as practical!

The greatest of the Great Houses is Pueblo Bonito. And astronomy is built into its very foundations. One wall runs through its centre, running exactly north-south, while part of its outer wall lies on a precise east-west line. But there's more to it than that.

Outside the southeast corner of Pueblo Bonito, a stone box is marked with an engraving of steps. GB Cornucopia explains that these match the stepped outline of the cliffs. A Sun Chief stationed here could have ticked off the months of the autumn, as he saw the Sun move along the horizon and set behind each step in the cliffs in turn.

ABOVE *A huge circular room – the Great Kiva – lies at the core of Pueblo Bonito, one of Chaco Canyon's Great Houses. Hundreds of people probably gathered here to celebrate the Winter Solstice, warmed by fires under a wooden roof that originally covered the entire kiva.*

The Sun reached the last step on October 29. On that day, the setting Sun began to shine through an unusual diagonally placed window on the first floor of Pueblo Bonito, casting a patch of light that gradually moved across the opposite wall from one day's sunset to the next. Eventually, it came to touch a protruding buttress – precisely at sunset on Midwinter's Day.

Perhaps the Sun Chief came indoors to this special room at the end of October, to make his observations sheltered from the freezing cold. 'The winter solstice is a really difficult time to be here,' says Cornucopia. 'It's very cold – our historical low in the canyon is -36°F (-38°C); it averages 9°F (-13°C)

Some modern astronomers have suggested that Chacoan skygazing went even further. Dominating one end of the valley, the massive Fajada Butte has an intriguing set of rocks, arranged so that sunlight falls between them in the shape of short knives. These 'sun daggers' shine onto spiral-shaped petroglyphs on both Midsummer's Day and Midwinter – and at the equinoxes too.

At the other end of the canyon, a pictograph high up on a cliff depicts a circular object with a fluffy tail, along with a crescent Moon plus a bright star. These just might depict Halley's Comet, during its visit in AD 1066; and a supernova that blazed forth in the sky in AD 1054, leaving as its remains a cloud of glowing gas we call the Crab Nebula.

'At one level I'm fascinated and I wonder is it an eye-witness account of the supernova of 1054,' says Cornucopia, 'but the evidence is circumstantial at best. On the other level, I don't care. It hardly matters when we think that whatever we read it to be, this painting tells us they were interested in the sky, they were affected by what went on in the sky.'

Because for them the sky wasn't something separate from the Earth. According to Clive Ruggles, the world's first professor of archaeoastronomy – a mouthful of a word for the archaeological investigation of our stargazing forebears – 'the sky was massively important to our ancestors, and so was a whole load of other stuff about what they saw in the landscape around them.'

Today, we artificially divide our surroundings into sky, landscape, plants, animals and so on. But, in the beginning, everything was part of a whole environment. 'Generally people in indigenous cultures in the past tried to make sense of the Cosmos – of the world around them – by drawing links between things,' continues Ruggles, 'things in the sky, things around them in the landscape and social things too – all mixed in.'

And as evening drew in on Chaco Canyon, we were treated to the skygazing Ranger spinning his stories of astronomy past and present to an enthralled bunch of campers. The crescent Moon was setting, while the Milky Way arched overhead. A meteor flashed through the sky. You could understand how the people living here felt so close to the heavens.

While the oral traditions of the Pueblo people have helped to flesh out ancient Chacoan skygazing, that's the exception when it comes to trying to understand the very roots of astronomy.

And the oldest, most impressive and most famous monument to the history of astronomy is also the most inscrutable. Standing isolated on the chalk plains of southern England, age and mystery seep from the massive stones of Stonehenge.

Several years ago, we were lucky enough to spend the night within the open walls of the great monument – on the night before the summer solstice.

The circle of great stones seemed to pull the stars downwards, forming a planetarium dome above us. We could have been keeping vigil with our distant ancestors who erected these stones over 4000 years ago – except for the music drifting over from the hippies' tent city across the road, accompanied by fragrant smoke.

As the sky lightened, a procession of white-robed Druids circled Stonehenge. To the northeast, a brighter glow coalesced on the horizon. The golden disc of the Sun emerged, immediately above the dark silhouette of the outlying Heel Stone.

Millennia ago, the British version of Chaco's Sun Chiefs were probably presiding exactly where we sat, celebrating the beginning of the longest day of the year. The Druids and the hippies certainly have no doubt of that, as they invade the site for their own rituals, including infant baptisms at the ancient Heel Stone.

OPPOSITE *Stonehenge is most likely an observatory for following the Sun through the seasons.*

BELOW *The ancient megaliths of Stonehenge, on the chalk uplands of southern England, have perplexed people for centuries. Why did our ancestors shape these giant stones to fit together in massive arches that have survived for four millennia? The answer may lie in the sky.*

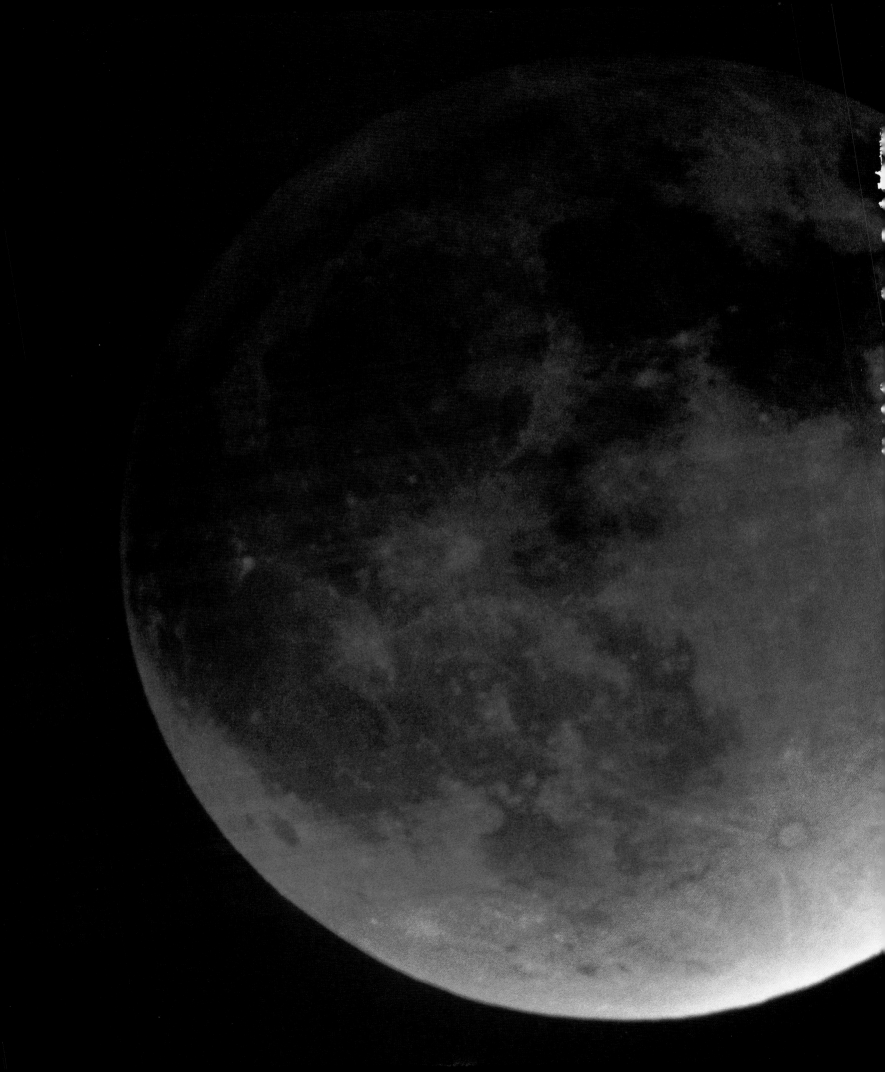

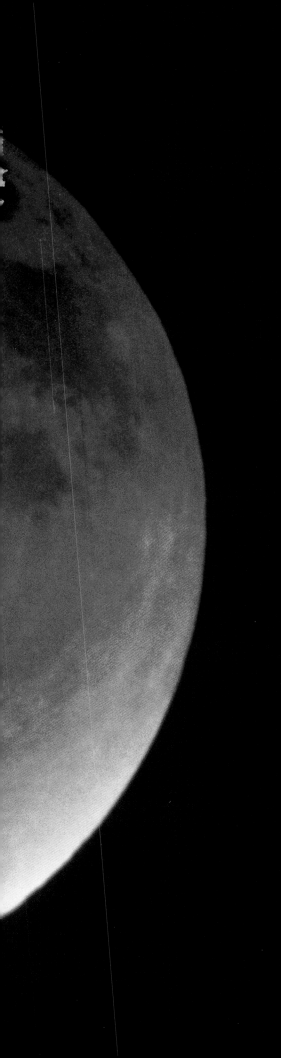

ABOVE *By night, Stonehenge has the aura of a planetarium – seeming to bring the stars down to Earth.*

LEFT *A total lunar eclipse is an awe-inspiring sight, when the Moon may turn blood-red. According to one theory, Stonehenge was a computer for predicting eclipses.*

Or were the Sun Chiefs standing at the Heel Stone, looking towards Stonehenge – and at the opposite end of the year? 'There's a lot of uncertainty about whether it's the direction looking out towards the Midsummer sunrise,' explains Clive Ruggles, 'or the direction coming in – towards the Midwinter sunset – that might have been the more significant.'

Either way, Stonehenge is a monumental shrine to the Sun. It was built from stones weighing up to 55 tons (50 tonnes), which have been shaped to fit together with precision joints, like carpentry on a massive scale. Some of the smaller stones have been transported from Wales, 135 miles (215 km) away.

'I think that Stonehenge was a huge temple,' declares Clive Ruggles, 'and I have no doubt that the solsticial alignment was part of its significance. By building this powerful monument that includes stones from exotic places, it puts Stonehenge very forcefully in the Cosmos as the centre of things.'

Is this all? In the 1960s, British-born astronomer Gerald Hawkins harnessed the power of the newly invented computer to unravel the mysteries of Stonehenge.

He concluded it was not so much a religious site as a giant observatory and computer rolled into one.

By sighting between the stones on the site, Hawkins deduced that early astronomers made precision observations of the Sun and Moon. He also suggested that a set of 56 holes surrounding Stonehenge was a primitive computer. The great astrophysicist Fred Hoyle improved Hawkins' model. He showed that ancient astronomers could have moved a set of four markers around the holes, to calculate eclipses of the Sun and Moon.

But Clive Ruggles is not impressed. He points out that Stonehenge was not built all at once, according to a single plan: its construction lasted over a millennium, the work of many different generations with perhaps constrasting agendas. And as you look at the history of the great monument, some of Hawkins' and Hoyle's ideas begin to fall apart.

In fact, the very earliest version of Stonehenge had nothing to do with stones. Around 2950 BC it was 'Earthhenge.' The site was merely a large circular ditch, with an entrance that led towards the Midsummer Sunrise.

Then it became 'Timberhenge.' Inside the ditch, a ring of 56 great wooden posts was erected; these have now rotted away to leave just a set of holes. This new evidence is bad news for anyone who wants to see Stonehenge as a computer. 'Most archaeologists would say that these holes held posts,' Ruggles explains, 'and these timber posts rotted away. That scotches the idea that they were open, and used for moving markers around.'

In addition, archaeologists have found rings of holes at other sites around Britain that number from 19 up to 100-plus. 'So the fact that there are 56 at Stonehenge I don't think is significant,' Ruggles continues. 'I don't think people were predicting eclipses at Stonehenge. But of course we can't say they weren't. And to be fair to Fred Hoyle, all he ever said was: if I were trying to use Stonehenge to predict eclipses, this is what I would do.'

But it was perhaps around this period – about 2700 BC – that the early British began to take the sky more seriously… 'People came along and made little offerings,' explains Ruggles. 'Bits of animals, sometimes a whole dog skull, placed very carefully.

And later on, cremated animals and even human cremations. They didn't put them all around the place – there's a big interest in the northeast and southeast directions, which to correspond to the most northerly rising and setting points of the Moon.'

Ruggles thinks that the first set of small stones – the Heel Stone and four other outliers – were then set up to help astronomers observe the Sun and the Moon. And in 2002, German archaeologists revealed a unique bronze disc that indicates our Bronze Age ancestors were keeping track of the sky in just this way.

Dating from 1600 BC – a few centuries later than Stonehenge – the Nebra sky disc is a little over 12 inches (32 cm) across, and adorned with astronomical symbols in gold leaf. A round circle is probably the Sun, while a crescent must be the Moon. Another curved shape may represent a boat that carries the Sun across the sky. Small dots are stars, with a group of seven stars probably showing the Pleiades star cluster, which we call the Seven Sisters.

Most interesting are the two golden arcs that stretch 82 degrees around the circumference, on opposite sides of the disc. This is exactly the extent that the Sun moves around the horizon, from Midsummer to Midwinter sunrise, as observed from central Germany. The arcs show that the skywatchers of Nebra watched the Sun's progression around the horizon with some precision.

Elsewhere in Europe, local Sun Chiefs may have used lines of standing stones to follow the Sun and Moon. We find spectacular megalithic alignments at Callanish in the Hebrides, and Carnac in Brittany. But at Stonehenge – around 2400 BC – people had grander plans. First they brought strange blue-coloured stones all the way from

BELOW *Sometimes called 'The Stonehenge of the North', the stone avenues of Callanish, on the Outer Hebrides off the coast of Scotland, are aligned roughly north, south, east and west.*

ABOVE *On the Midwinter Solstice, the Sun rises behind the leftmost tower of the 13 that make up the newly discovered solar observatory of Chankillo, in Peru. During the next six months, it rises behind all the others in turn.*

Wales. Then they shaped massive boulders, and built them up into giant portals and lintels of stone, and Stonehenge assumed its present imposing structure.

While Ruggles accepts that the earlier stones could have been set up to observe the Sun and Moon, he scratches his head over Gerald Hawkins' idea that all these massive stones are set up as astronomical markers: 'You've got many pairs of stones, and you can always find alignments if you take enough pairs of things.'

And the sheer scale of Stonehenge raises doubts that it was built merely for astronomers to forecast the sky and the seasons.

'Well to assume that is rather like saying you need to put the US economy onto a full budget for a hundred years to work out your shopping list,' retorts Allan Chapman, a leading historian of science at the University of Oxford. 'You don't need all of that colossal investment in stone, earth-moving, all the rest of it, just to know when to put the pigs out or when to gather in the wheat!'

Ed Krupp at the Griffith Observatory expands. 'Usually monumental architecture isn't dedicated to observing the sky. It's a way that people express how they feel at home in the Universe; but also how they control the Universe.'

When people lived as nomads, giant monuments were out of the question. The first chiefs or kings of farming communities, however, wanted to create a symbol of their power, which in turn would co-ordinate the community.

ABOVE *Built around 3200 BC, the great tomb of Newgrange in Ireland has recently been restored to its pristine condition. For a few mornings around Midwinter's Day, the rising Sun shines straight up the long passage within Newgrange, dramatically lighting up the central tomb.*

Across the Irish Sea from Stonehenge, we can see that celestial order integrated into the huge funerary mound of Newgrange, in Ireland's Boyne Valley. While Stonehenge was still a twinkle in someone's eye, a major chief was buried in Newgrange at the end of a long passage grave. The stone-lined corridor faces southeast, so the rising Sun shines directly up the long passage on Midwinter's Day.

And on Jersey, one of the Channel Islands between England and France, the great mound of La Hougue Bie has perpetuated its religious symbolism from the earliest times. Right at its base is an ancient passage grave that admits the Sun's first rays on the morning of the Spring Equinox. And the mound is crowned by a medieval Christian chapel.

'People use the sky for that kind of co-ordination,' Krupp continues. 'After all, the sky is the fundamental regulator – it provides the basic references for time, for direction. This is the order of the world. People then incorporate it into their architecture, into their rituals, into their costumes, into every aspect of life to demonstrate that they are integrated with that order.'

The most convincing example of ancient people using the Sun to regulate their yearly calendar came to light in March 2007, when Clive Ruggles and his colleague Ivan Ghezzi published their interpretation of the Thirteen Towers of Chankillo. These are a baker's dozen of stone piles that rise from an elongated hilltop in Peru, like the spines on a dinosaur's back. They date from AD 200 – many centuries before the Inca civilization that we usually associate with this region.

'This is an extremely exciting site,' enthuses archaeoastronomy guru Ruggles. 'It's the clearest example I've ever seen of a solar alignment.' At the winter solstice, the Sun rises behind the leftmost tower in the row. The progress of the months is marked by the Sun rising behind each of the other towers in turn, until six months later it is rising behind the rightmost stone edifice.

But the fusion of sky, Earth and ourselves doesn't necessarily require showy monuments. For some peoples, it resides deep inside themselves. And in Australia, we can find a unique window into the astronomy of the past. In a culture that makes no distinction between the past and the present, ancient oral traditions of the skies have passed down to the present day.

'The Aboriginal cultures are all about integration,' explains Ray Norris – a British radio astronomer who now works in Australia and has been seduced by the lore of the native skywatchers. 'There's no clear distinction between the sky and the land, or between what's magic and what's everyday.'

And even the origin of the Universe becomes one with everyday life. The Dreaming was when the creator spirits formed the world; but it's also a process that's going on now. 'Most creator spirits are actually still around,' continues Norris, 'everything from little meany spirits which are like elves, right up to the big creator spirits who may live in places like Uluru.' The planet Venus is one of the creator spirits. To the Yolngu people, living in Arnhem Land in the north of Australia, she is called Barumbir.

BELOW *On the spring and autumn equinoxes, the rising Sun shines through the entrance of La Hougue Bie, in Jersey, and up the passageway within. This artificial mound has held a special significance for over 5000 years; in medieval times, a chapel was built on top.*

LEFT *The skies above Australia provide both a calendar and spiritual guidance. In one legend, the stars of the Southern Cross (top centre) represent the eyes of the god of death and of the first man to die.*

BELOW *Uluru (formerly known as Ayer's Rock) has a special significance to the Aboriginal people of Australia, as the home of powerful creator spirits.*

'She comes in from the sea,' explains Norris, 'from the east and travels westwards across the land. And as Barumbir walks, she's naming objects and animals, saying 'this is where you fish,' 'I'm going to make sand-dunes here' or 'these are kangaroos, eat these.' And there's a song she sings which brings the land into being.'

And the songs are repeated to this day. Norris recalls sitting on a beach in Darwin, eating fish and chips, when some Aboriginal kids in baseball caps and tee-shirts started singing: 'It's not some Eminem rap. Just to amuse themselves, they're singing the Yolngu songs. And it's just a lovely sound.'

The present-day Yolngu also use Venus as a way of communicating with the spirits in the sky. 'They have this wonderful Morning Star ceremony,' Norris tells us. 'It starts the previous evening, at sunset, and it involves poles with feathers on, which represent the lady star Barumbir. As she rises in the morning, this is the creative spirit actually coming towards them.' With brilliant Venus holding open the doorway to heaven, the Yolngu can speak to their ancestors, or relations who have recently died, and keep them in touch with all the news – 'like little Johnny's grown his front teeth,' adds Norris.

But there's no single myth for Venus, or any of the other celestial sights. There are literally hundreds of Aboriginal groups across the continent, each speaking a different language and holding different traditions. Interestingly, they generally view the Sun as being female, and the Moon as a male – the opposite to most cultures in the world, which have a Sun-god and a Moon-goddess.

Some of the Aboriginal stories may date back the time when these people arrived in Australia over 40,000 years ago. They would then be the very earliest astronomical traditions in the world. But in the land of the Dreaming, it's impossible to know how far back the story-weaving began.

When it comes to Venus, for instance, the Aborigines of Ooldea, on the edge of the desert in South Australia, have another account of the – male – Morning Star. They tell the story of a young couple, who were traveling with an older man acting as their chaperone. When he kept on stopping them from doing what they wanted to do at night, they tricked him into climbing a tree – and then chopped it down, to leave him stranded in the sky. But he became the Morning Star, and is now always keeping an eye on anyone who is tempted to stray.`

That's not the only example of the sky providing a reminder of the Aborigines' moral code. The people living near the Clarence River in New South Wales point out a red star that warns of the dangers of adultery. In the western world, we call it Aldebaran – the eye of Taurus, the bull – but here it is a man called Karambal. He seduced a young woman, but unfortunately she was already married to a great warrior. Karambal escaped by hiding in a tree, but the wronged husband set fire to it, sending Karambal up to the sky where he still burns today.

The constellation of Orion – seen in the West as a mighty hunter – also has a different interpretation in Australia. That's partly because the stars are seen 'upside down' from the southern hemisphere. Contemporary Australians often pick out the central three stars of Orion's 'Belt' and 'Sword' as rather nicely defined Saucepan!

'The Yolngu see the three stars of Orion's Belt as three men sitting in a canoe,' says Ray Norris, returning to the more ancient traditions, 'with Betelgeuse and Rigel as the front and back of the canoe.' Orion's Sword – comprising fainter stars and the glowing patch of the Orion Nebula – is a fish caught on a fishing line. Norris enthuses: 'Once you've been told this, and you've seen it, it's actually really dramatic.'

But the Aboriginal constellations aren't all made up of stars strung together. While the western star-patterns are like joint-the-dots puzzles, the indigenous people of Australia have other ways of seeing the sky.

You can understand why, when you look up on a moonless night in the Outback. The sky is so dark, and so clear, that it's overflowing with stars, filling almost every spot of the heavens. Amongst the starry multitude, your eye picks out two things. First, you see the fuzzy patches where stars assemble into clusters. Then you start to notice the dark patches, where there aren't any stars… The Milky Way is a wide glowing band. But along its length, dark clouds block the light from stars behind. To many Aboriginal peoples, these clouds are the silhouette of a giant emu. The head of the celestial bird is a small black patch, which westerners call the Coal Sack, lying next to the Southern Cross. Its body runs down into the western star pattern of Scorpius (the scorpion).

ABOVE The creation ancestor Namondjok dominates this rock painting in the Kakadu National Park, in northern Australia. According to local legends, he broke kinship laws through incest with his sister, and was banished to the sky. He can now be found as a dark patch silhouetted against the Milky Way. The smaller figure beside him is Namarrgo, the Lightning Man; while below Namondjok we find the Lightning Man's wife, Barrginj. Their children are blue-and-orange grasshoppers, which appear just before the wet season and call to their father to conjure up his thunderstorms. These figures may date back thousands of years, but have been continually repainted. They demonstrate the seamlessness of the Aboriginal people's concept of sky, land, weather, animals and moral values – all part of the Dreaming.

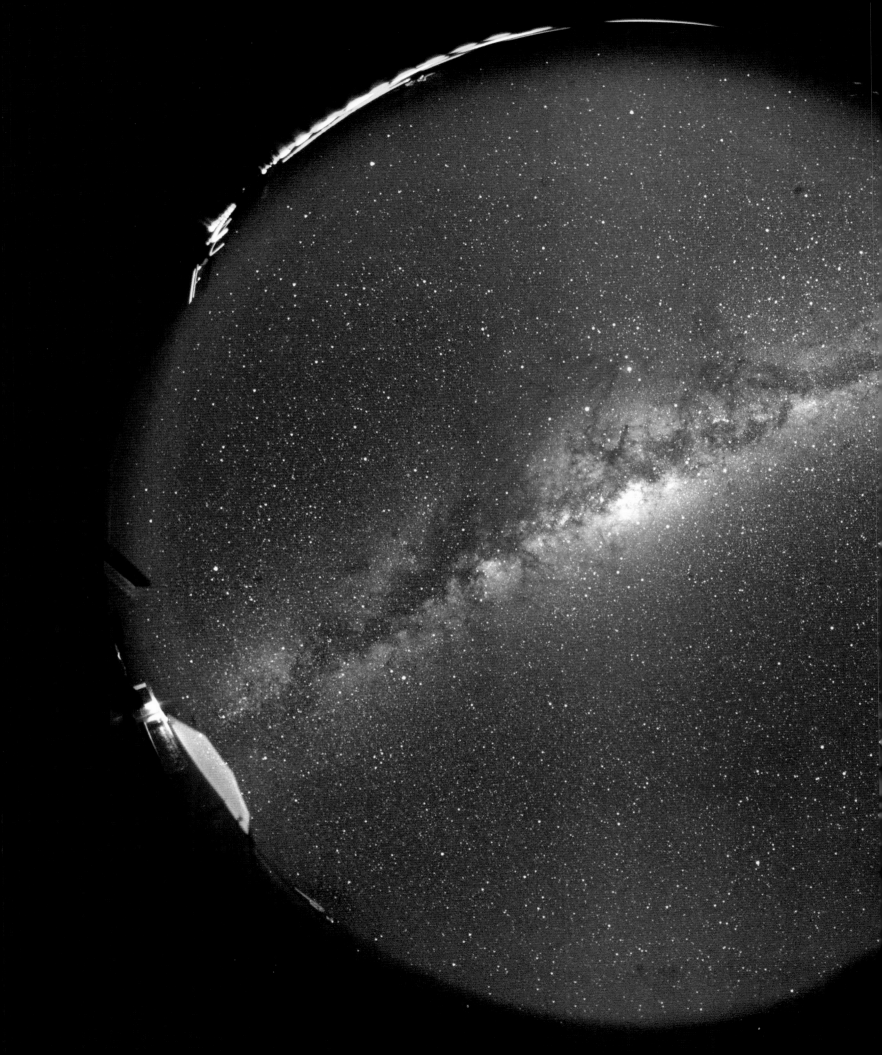

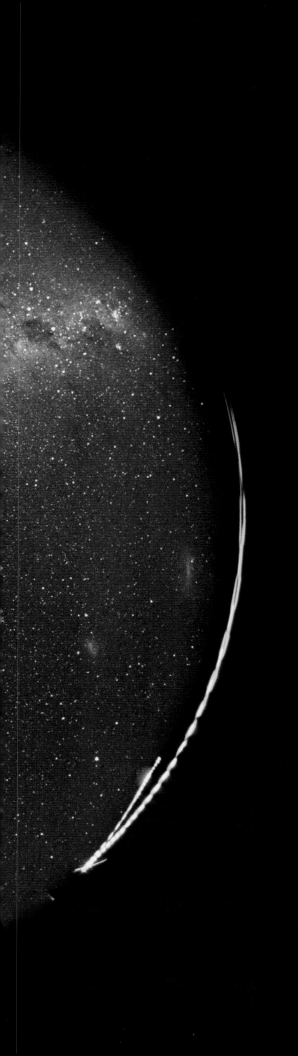

Near Sydney, there's an ancient carving in the rock, depicting the outline of an emu – but with its legs oddly splayed out behind. 'What's interesting,' muses Norris, 'is that this engraving doesn't look like the emus that run around – but actually looks more like the emu in the sky.'

And there may be a practical aspect to this engraving. In May, the emu in the sky stands directly over the carved emu on the ground – marking the time for the Aboriginal people to collect emu eggs.

Other signs in the sky also helped the indigenous Australians to find food in their barren lands. The 'mallee-fowl' constellation (Lyra) showed the time of year to harvest the eggs from these hen-like birds. And for people in the western deserts, the season for collecting tasty young dingo pups was signaled by the appearance of the Pleiades.

This star cluster has excited people all around the world, from the earliest times. In the western world today, the Pleiades are largely hidden from view by the light pollution that makes the sky one vast glowing dome.

But in the desert, or at sea, the Pleiades are just about the first thing you home in on. This bright, compact group of stars is unlike anything else in the sky. Throughout the world, people have identified its stars as a group of young girls – hence the Seven Sisters – often being pursued by an aggressive male, who may be the red star Aldebaran or the constellation Orion.

Across the Pacific from Australia, people in the Andes regularly observed the Pleiades to ensure a bountiful supply of food. They were not hunters, though, but farmers. If the Seven Sisters were clear and bright, they would plant their potatoes in October. But they would delay the planting if the Pleiades were dim and hazy.

Modern science has recently caught up with the wisdom of the Andean ancients. When the Pleiades are dim, they're being hidden by a veil of high-altitude cloud, associated with an El Niño in the Pacific. This change in the ocean's circulation leads to a drought in South America, and would spell ruin for the potato crop.

In the midst of the vast expanses of the Pacific, the Polynesians used the Pleiades as the basis of their calendar. Not for them the fixation with the Midsummer and Midwinter Sun, that we find at Stonehenge or Chaco Canyon. The Polynesians divided their year into two halves. The first started in December, when they saw the Pleiades rising as the Sun set; while the second half of the year began in May or June, when the Pleiades were first visible in the morning sky. And the Polynesians had an intimate relation with the sky for another reason. They relied on the stars to navigate, as they sailed across thousands of miles of the empty Pacific Ocean.

LEFT *Dark clouds along the Milky Way form the Aboriginal constellation of the Emu. As seen in this all-sky view, its body lies in the centre and its head at top right.*

'From the central Eastern Polynesian islands,' explains Ben Finney, an anthropologist from Hawaii, 'canoes sailed north across the equator to distant Hawaii, southeast to tiny Easter Island off the coast of Chile, and back southwest to the great islands of New Zealand – to complete the settlement of an island realm spread over an area almost as large as Europe and Asia combined.'

To check how they accomplished this almost miraculous feat, Finney has built replicas of the Polynesians' voyaging canoes, the huge double-hulled vessels in which they sailed, with their families and crops and animals, to colonise their new worlds.

'The first islands they came to were already inhabited, but soon every island they came to was uninhabited,' Finney continues. 'And they kept pushing and pushing, because they were the only people in that part of the world. Every island was new and virgin, and open to them.'

Finney's first canoe was called Hokule'a, the Polynesian name for the star Arcturus. In 1976, he sailed with a native navigator from Hawaii all the way to Tahiti; and then back with a young Hawaiian who had learnt the traditional methods of star-navigation, Nainoa Thompson.

Standing on the deck of Hokule'a – even in Honolulu Bay – you feel light years away from the bright lights, crowds and noise of Waikiki. The crew of a dozen volunteers is bonded like family, much as the ancient Polynesians must have done to survive the long voyages. A magical moment comes when a school of dolphins cavorts around the twin bows. Above all there's a sense of peace; of being at one with both the ocean and the stars above.

Nainoa is our guide to both. In some Polynesian islands, there's no such word as 'astronomer' – if you want to find your way around the sky, you ask for a 'navigator.'

'We use the stars in a number of ways,' he explains. 'One is to hold direction: the places where the stars rise and set give us directions we can use to tell the bearing of the canoe.'

These directions are summed up in the ancient Polynesian 'star-compass.' It divided the horizon up into 32 different directions, corresponding to the rising or setting of individual bright stars and the all-important Pleiades. But the star-compass wasn't taken to sea. In fact, it didn't physically exist at all. A Polynesian navigator like Nainoa kept the star-compass in his head.

An early Pacific navigator also relied on the stars to check how far the canoe was north or south of the equator. As they traveled south from Hawaii, round the curve of the world, the Pole Star to the north gradually sank lower and lower. South of the equator, it disappeared altogether, and they had to rely on the stars of the Southern Cross.

The Polynesian navigators also knew that different stars passed directly overhead on the various islands they called home. For Hawaii, it was the orange star Arcturus. When they sailed southwards to Tahiti, Arcturus gradually toppled from the zenith. Meanwhile, brilliant white Sirius would rise higher.

When Sirius was overhead, the navigator knew they had reached the latitude of Tahiti. Now it was a matter of turning the great canoe east or west, and keeping a look-

RIGHT *The ancient Polynesians crossed the Pacific Ocean in giant double-hulled voyaging canoes – like this recent exact replica – carrying their families, along with plants and animals for colonizing newly discovered islands. Without any sophisticated instruments, Polynesian navigators charted a safe course across thousands of miles of empty sea by relying on the stars.*

out for the signs of land – birds, driftwood and towering clouds – that would lead them to a precise landfall.

Without any accurate instruments, the Polynesians' close relationship with the sky made them as at home in the wide ocean as most of us are on land. It was a feat that astounded the European navigators of the sixteenth century, when they finally developed the technology that allowed them to trespass among the scattered islands of the Pacific.

'The Polynesians didn't regard the ocean as an alien environment,' explains Finney, 'but one that was utterly natural – and essential to spread of human life.'

And spread human life they did, over the only portion of the globe that had so far been free of humankind. The Polynesians' final port of call was New Zealand, where the people known as Maori arrived around AD 1000.

In the wake of the great Pacific voyages, Homo sapiens reached every habitable region of our planet. And wherever we find humankind, we find a fascination with the sky. People have an inbuilt urge to relate to the heavens. Under the familiar stars, we don't feel alone at night. The rhythms of the Sun and Moon provide a pattern to our lives. Farmers have instinctively tuned into the revolving heavens to timetable their crops; hunters pick up cues to the most rewarding seasons; and navigators can unerringly traverse the Earth from the great map of the skies above.

And perhaps there's a deeper link, too. The harmonies of the heavens contrast with the chaos and unpredictability of everyday life – with its share of sadness, unfairness and cruelty. The sky is an inspiration.

The words of a Maori saying ring through from the earliest days of skywatching: 'The Sun, Moon and stars all live together peacefully without quarrelling. Evil is unknown to them. When can we become like them?'

LEFT *A meteor streaks past the Pleiades star cluster, in this stunning view from the Joshua Tree National Park in California. The Pleiades – also known as the Seven Sisters – crop up in sky legends from every corner of the planet. Navigators used them to cross the oceans; farmers relied on the Pleiades to keep track of the seasons; and the peoples of South America even found that the appearance of the Pleiades could predict the extreme weather conditions of El Niño.*

Reading the Heavens

'Fifty-second day: fog until next dawn.
Three flames ate the Sun, and big stars were seen.'

So runs the oldest written account in the history of astronomy. And, amazingly, we can date it precisely – to June 5, 1302 BC. To anyone who's witnessed the incredible sight of a total eclipse of the Sun, the description strikes a chord even today. The 'three flames' are streamers in the Sun's atmosphere, the corona, reaching out from around the black orb of the Sun's eclipsed disc. And when the sky goes dark, we suddenly get to see 'big stars' – in this case, Sirius and the stars of Orion, along with the planets Saturn and brilliant Venus.

This eclipse was noted on an ancient piece of turtle shell, found near the city of Anyang in central China. It's an ancient 'oracle bone,' used for foretelling the future back in the Shang dynasty, around 3000 years ago.

'They were turtle shells, pieces of ox bone, things like that,' explains Richard Stephenson, a British scientist who taught himself ancient Chinese and is now an expert on the astronomy of the Far East. 'They would inscribe a question to the oracle – i.e. the ancestors – on one side of the bone. Then they would push in a hot needle from the other side, and they would interpret the direction of the resulting cracks, whether the answer was basically yes or no.'

In this case, the question is: 'Diviner Ge asks if the following day would be sunny or not.' The other side records the outcome – that the sunlight was rudely interrupted by the Moon!

In 1989, researchers at NASA's Jet Propulsion Laboratory pinned down the date of this oracle bone, by checking out what eclipses would have been visible during the ancient Shang dynasty, and on 'day 52' of the 60-day Chinese calendar cycle.

RIGHT *A traditional Chinese lion – his paw on a globe of the world – guards a gateway within the Forbidden City in Beijing. The Emperors of China set up a bureau of astronomers to report every odd happening in the sky.*

OPPOSITE *Chinese astronomer Zu Chongzhi, drawing with a pair of compasses. In the fifth century AD, Zu proposed a new calendar, accurate to better than a minute in the year. He was building on a tradition of Chinese astronomy dating back almost two millennia before his time.*

ABOVE Massive astronomical instruments stand exposed to the elements at Beijing's Ancient Observatory. Built in the seventeenth century, the bronze armillary sphere (left) and the quadrant combine millennia-old Chinese astronomical tradition with European ideas imported by missionaries: apart from the dragons, they bear more than a passing resemblance to the instruments constructed by the Danish astronomer Tycho Brahe (see Chapter 5).

With ancient Chinese records like these oracle bones, the history of astronomy moves into a new phase. No longer are people content to just 'live with the sky,' working out roughly when to plant crops or to hunt for young animals. This was the start of precision astronomy.

What drove the change, ironically enough, wasn't science in the modern sense of the word. The ancient Chinese weren't particularly interested in what was in the heavens, nor why eclipses happened. They had more immediate concerns.

For them, the sky was the mirror of the Earth. By keeping a close eye on the heavens, the Emperor could check what was happening in China. Generally the celestial bodies moved in a regular way: the Sun following its yearly progression, the Moon waxing and waning every month and the planets moving in a slow and stately progression.

But sometimes the heavenly order broke down. A new star might blaze out. Or a comet – a 'broom star' to the Chinese – might trample through the star patterns. Aurorae could light up the night sky; or dark spots blemish the face of the Sun.

Throughout Chinese history, the country's rulers kept an anxious eye on the sky. But things became more formalized after Qin Shi Huang united the warring states in 221 BC. The First Emperor founded the Great Wall, and was buried near Xi'an with his famous Terracotta Army. One ancient account says that Qin's own tomb contains a fabulous double map of the land and sky: 'Below was the map of the Earth. The 'hundred rivers' of the Empire were modelled in mercury: some machines made it flow and conveyed it to and fro. Above everything was the starry vault.'

Shortly afterwards, an Astronomical Bureau was established. It has to count as one of the longest running bureaucracies in the world, operating for over 2000 years.

RIGHT *The observatory's celestial globe is studded with stars, and mechanically driven to show them rising and setting, in sync with the real sky.*

And it has provided us with the longest continuous set of sky records – a priceless archive that's been mined by scholars like Richard Stephenson to track down records of skysights from long ago.

The Ancient Observatory of Beijing stands to this day, next to the city's busy ring road and overshadowed by office blocks bearing the names of the world's leading electronics companies. Look up from the roadside, and on top of the squat tower you'll see the skeletal outlines of bronze instruments. These mainly date from the seventeenth century, and are a curious hybrid of the designs of the Danish astronomer Tycho Brahe with indigenous Chinese ideas, such as the writhing dragons that form the supports.

For centuries before this, a team of five astronomers would keep an eye on the sky every night from such 'Platforms of Star Watching.' One would look out to each of the four compass points; while the fifth would lie back and look straight up (surely the best job!).

If an astronomer spotted something that was not expected – perhaps a comet, or a 'guest star' – he would note exactly where it appeared. The next morning he would report it to the Astronomer Royal, who interpreted its meaning to the Emperor, the embodiment of China itself.

Yes, it was more astrology than astronomy. But at this time, people felt such a close affinity with the sky – they didn't know that celestial objects lie millions or billions of miles away – that it made sense to believe the sky was intimately related to events on Earth.

BELOW *An engraving from 1755 shows the Beijing Ancient Observatory looking scarcely any different from today – except that it's now overshadowed by office blocks!*

OPPOSITE *The Suzhou star chart was engraved in stone in AD 1247. Its 1434 stars are grouped into 283 small constellations. The only patterns familiar to western eyes are the Plough (Big Dipper) – back-to-front near the centre – and Orion (at the "four o'clock" position).*

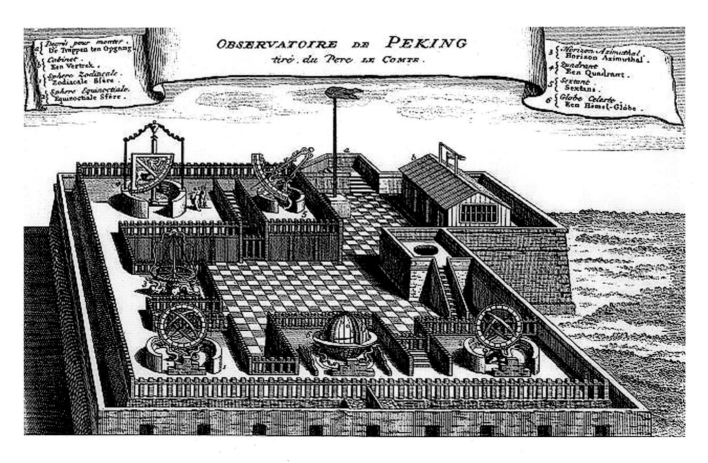

OBSERVATOIRE DE PEKING
tiré du Pere le Comte.

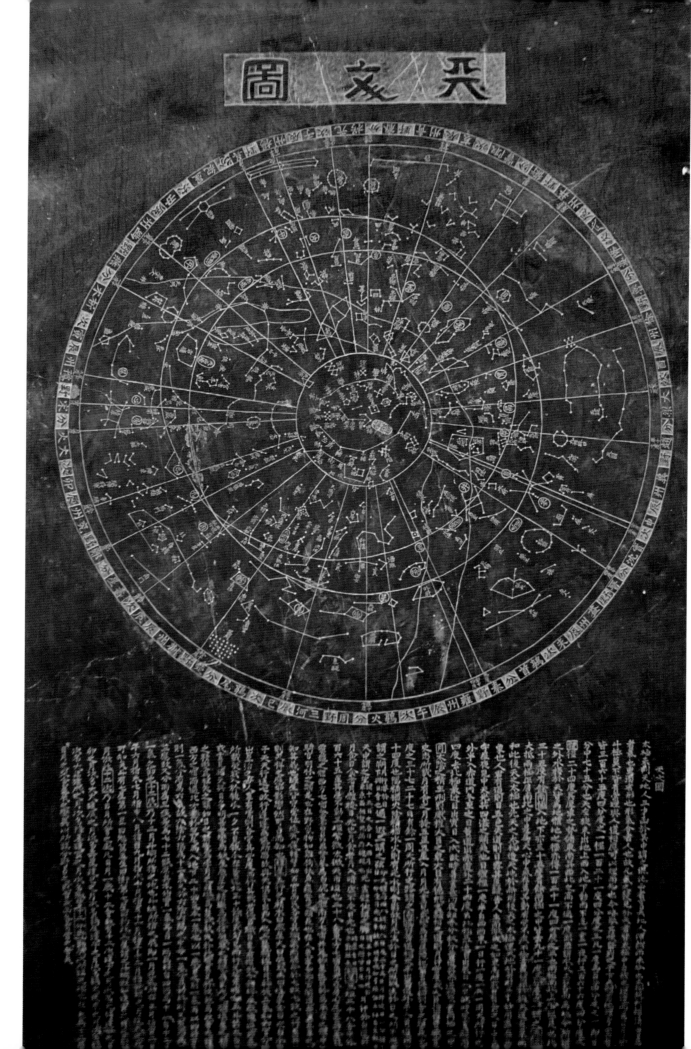

'Right from as early as we can trace,' declares Richard Stephenson, 'the main emphasis was astrology. And without astrology we would have virtually no astronomical records from East Asia, because that was the whole reason for watching the sky.'

But this wasn't Chinese astrology of the kind we know today: where fortune cookies will tell you what will happen tomorrow if you're born in the Year of the Rat. Deep in China's history, no-one thought that the heavens related the fortunes of ordinary individuals. The sky's influence was at a national level.

'Certainly by around 200 BC,' Stephenson continues, 'this idea of political astrology had grown up. Various events in the sky could affect the country as whole, and particularly the Emperor and the ruling family.'

On December 7, AD 185, the skywatchers were perturbed by a brilliant object low in the sky. They reported: 'A guest star appeared within the Southern Gate. It was as large as half a mat; it showed the five colours and it scintillated.'

On reading this account, Richard Stephenson recognized the Southern Gate as twin bright stars we now call Alpha Centauri and Beta Centauri. He suggested the 'guest star' was a supernova – a dying star that has blown itself apart in a brilliant explosion. In 2006, astronomers using the Chandra observatory in orbit around the Earth identified a superhot cloud of gas which is the fireball left after this 2000-year-old explosion.

The Astronomer Royal of the time, though, had more terrestrial violence in mind. The new star had upset the harmony of the heavens, indicating that something was very amiss in China. Indeed, the official history relates the result: 'The governor of the metropolitan region Yuan Shao punished and eliminated the middle officials. Wu Kuang attacked and killed He Miao, the general of chariots and cavalry, and several thousand people were killed.'

To identify exactly where trouble was expected – or who was involved – the Chinese divided the sky up into many small constellations. Whereas astronomers in the West have 88 star patterns, the ancient Chinese had 283 constellations.

Most of the Chinese constellations were small, often made up of only two or three stars, with names such as the Court Eunuchs and the Sombre Axe. There are very few celestial shapes that Western astronomers share with the Chinese. The exceptions include the Plough (or Big Dipper) - which in China was the Northern Lady - the stars of Orion, and the tail of Scorpius (the scorpion), which is a dragon's tail to Chinese eyes. 'But I defy anyone to try to identify anything else from the star map,' challenges Stephenson, 'it's really as difficult as that.'

In the West, for example, we are used to connecting up five brightish stars in the northern sky to make a distinctive W-shape, called Cassiopeia after a queen in ancient Greek legends. The Chinese didn't see the 'W'. For them, these stars made up three

LEFT *The Crab Nebula is the debris from a supernova that the Chinese saw explode in AD 1054.*

different constellations. Three stars represented a famous charioteer, Wangliang, with his horses. Another star was part of a line of fainter stars called Gedao, a path across the mountains (the Milky Way); while the other was Fulu – the Auxiliary Road.

Two supernovae appeared in this region – in AD 1181 and 1572 – and were duly noted. But the most important 'guest star' in the Chinese records dates from 1054. On July 4, the sky watchers reported a stunning sight: a star so bright that it was visible in daylight for 23 days. It lay in the constellation that Westerners know as Taurus; and its remains are visible today as a glowing gas cloud called the Crab Nebula. At the heart of the twisted remains of the old supernova lies a rapidly spinning relic of the explosion, the Crab Pulsar.

For today's astrophysicists trying to understand the nature of pulsars, the Chinese records provide one vital clue they can't find out any other way: the age of the Crab Pulsar. It formed when the core of the supernova collapsed, so the Chinese skywatchers reveal that the pulsar is just over 950 years old – a mere youngster on the cosmic stage.

'One of the key things is that you have a definite date,' Stephenson continues, 'so you know precisely how long the remnant has been evolving. Another point is that these historical records will often give you a good estimate of its brightness. In the case of a supernova in AD 1006, the Chinese said it was so bright you could see things on the ground by its light. And I think there's a third point: once you've given it a date, it's not just a reference number in a table, it's got its own signature, its own character.'

And it's not just supernovae that turn up in the Chinese records. In 240 BC, the skywatchers saw a 'broom star' that moved gradually from the east to the west. This is the earliest mention in the world of Halley's Comet.

A later account of the comet's return in AD 530 is much more detailed. It reads in part: 'On August 29 a broom star was seen in the morning at the northeast direction. Its length was six degrees and its colour pure white … On September 1 it was one degree to the northwest of Xiatai [a star in Ursa Major]; it went out of sight in the morning. On September 4 in the evening it reappeared in the northwest direction…'

From these observations, scientists have been able to calculate the comet's orbit with greater precision, helping European space controllers to send the *Giotto* spacecraft with unerring accuracy through the heart of Halley's Comet in 1986.

Eclipses were even more important to the Chinese. And the greater the amount of the Sun that was covered, the greater the calamity. 'They were regarded as very much affecting the Emperor,' explains Stephenson. 'There was a total eclipse in 181 BC, and the Emperor's dowager was really alarmed about it. A couple of years later, she died.'

Apart from its role in eclipses, Chinese interest in the Moon extended to using its night-to-night movement to divide up the sky into 28 wedge-shapes, the lunar mansions. The planets were not so important, though Jupiter had a special claim to fame. It was the Year Planet, and as Jupiter moved round the sky in the course of 12 years, it marked out the successive years that would later be named for the 12 different animals of Chinese astrology.

ABOVE *Early in 2007, Comet McNaught became the brightest 'broom star' in 20 years, visible at its best from the southern hemisphere. A spectacular comet like this was guaranteed to feature in ancient Chinese records.*

Perhaps because they weren't always trying to pin down the planets' motions to some celestial machinery, the Chinese also had a surprisingly modern view of cosmology. 'The Sun, Moon and the company of stars float in the empty space,' wrote one scholar, and another added: 'each has its own course, like the tides and waves of the sea and rivers and the movements of the numerous living creatures.'

Around the world, in the steaming jungles of Mexico, astronomers in the Mayan civilization were not happy just to let these cosmic creatures move so freely. In particular, they had their eye on Venus.

The most brilliant object in the sky, after the Sun and the Moon, Venus never strays far from the Sun. We see it hanging in the twilight sky like a tiny lantern, either after sunset as the Evening Star, or before dawn in the guise of the Morning Star.

Mayan buildings rise in the forest canopy of the Yucatan like prehistoric monsters. And they are almost as mute. The Spanish missionaries who followed the Conquistadores in 1519 had no interest in the natives' pagan culture. Incredible as it seems to us today, they systematically gathered up and burnt all of the books that the Mayans possessed. A whole culture went up in smoke.

Except for four books made of bark – known as 'codices' – which somehow made their way to Europe unscathed. Most fascinating is the *Dresden Codex*. It reveals that the Mayans were not just primitive savages with a penchant for constructing huge edifices, but cutting edge astronomers and mathematicians.

As with all cultures at this stage, however, their interest was not so much scientific as astrological. Archaeoastronomer Clive Ruggles says: 'Their astronomy achieved huge sophistication, but the motivation was basically to prognosticate what was going to happen to the kings, and hence to the people.'

And the Mayans' complex maths moved them on one step from the Chinese. They didn't just respond to signs in the sky. At least as far as the Sun, Moon and planets were concerned, their astronomers could see what was coming. The *Dresden Codex* describes in detail how the planet Venus moves in the sky; and it includes predictions for its future motion that are accurate to one day in 500 years.

Some of the Mayan buildings have Venus programmed into their architecture. The Governor's House at Uxmal lines up with a distant pyramid marking the southernmost point where Venus rises on the horizon.

A strange cylindrical building at Chichen Itza – called the Caracol (the snail) – was clearly an observatory. Standing inside, a Mayan astronomer could peer out through long shafts in the wall that focused on particular portions of the horizon. Some of the sight-

ABOVE *The stepped pyramid El Castillo ('the castle') at Chichen Itza symbolizes the yearly calendar. Each of its four staircases has 91 steps, which together with the top platform add up to the 365 days of the year. And, at the equinoxes, the Sun casts a writhing snake-like shadow on the sides of the steps.*

lines at Caracol aim the astronomer's vision to the rising and setting of the Moon. Others line up with Venus's northern and southern rising points.

But was Venus really their main focus of attention? Astronomical historian Ed Krupp thinks we could be in danger of being misled by the fact that so little remains. 'The Sun was certainly important in this culture, and the Moon as well,' he argues. 'When you think about the fact that there are only four surviving Mayan codices, I don't know if they ignored the rest of the sky. Venus was perhaps no less and no more important than other key celestial objects.'

The Mayan astronomers had seriously cracked the problems of the calendar, as well. In fact, they ran three separate calendars at once. One repeated every 260 days, the second every 360 days, while the third kept pace with the year, with a length of 365 days. And their observations of the Moon meant that they could predict eclipses accurately.

The case of the Maya shows how difficult it is to work out the astronomical sophistication of a culture just from surviving buildings. Ruggles concludes: 'In the Mayan case, the fact that we happen to have these codices, like the *Dresden Codex*, shows what incredible detail they went into correlating the different cycles of eclipses and so on. There's actually nothing in the architecture that would really convince us of that.'

But there's one stunning piece of architecture at Chichen Itza which has a blatant relation to astronomy. At the heart of the ancient city stands a great pyramid, made of nine platforms of diminishing size stacked on top of each other.

A steep set of stairs runs up the middle of each side. And at the equinoxes, the setting Sun casts a shadow of the platform-edges onto the side of the northern staircase. It forms the shape of a giant chequered snake, that writhes towards a pair of carved snakes' heads at the base of the stairs.

Its counterpart in the Old World, the set of three huge pyramids at Giza in Egypt, are even better aligned to the four points of the compass – so accurately, that archaeologists argue to this day how their builders 4500 years ago could have got it so right.

The Great Pyramid of Khufu (sometimes known by his later Greek name, Cheops) is the oldest of the ancient Seven Wonders of the World; despite its seniority, it's the only one to survive to the present day.

And 'great' it certainly is. If you transported it to Manhattan, Khufu's pyramid would cover four city blocks, and rise to over one-third the height of the Empire State Building. When you stand under it, the pyramid seems to rise never-endingly to heaven. Inside, a spacious gallery slopes steeply upwards, between precision-smooth walls, towards an empty chamber, that's guarded by a cunningly counterbalanced massive stone door. Inside, the 'King's Chamber' contains nothing but a massive – empty – stone sarcophagus.

The people who built this pyramid worked to an amazing precision. Over the whole of the huge area it covers, its base is level to the thickness of your finger. And it's lined up north-south with a precision that astronomers didn't achieve again until the time of Tycho Brahe, 4000 years later. Yet we don't know if they divided up the sky as meticulously as this. 'The Egyptians had patterns of stars and these are recorded in a number of tomb sealings,' explains Harvard historian of astronomy Owen Gingerich, 'yet they never put the stars in those illustrations. So it is very difficult to know exactly what the giant hippopotamus represents – it may be fully a quarter of the sky.'

There was one star, though, that underpinned the whole of Egyptian life: Sirius. When the Egyptian skywatchers saw the brightest star rise in the morning sky just

before the Sun, they knew that the Nile was about to flood. This annual inundation covered the land with fresh soil, enabling them to grow another year's crops.

'I would suggest that their real concern with the sky was basically time-finding,' opines Allan Chapman, a leading historian of science at Oxford University.

As far as the progression of the year was concerned, the Egyptian calendar was to be found in the sky. In addition to magnificent Sirius, they picked out another 35 stars spread around the heavens, at roughly equal intervals. During the year, these chosen stars would rise before the Sun at roughly 10-day intervals – hence their name, the decans.

Once they'd picked these stars, the Egyptians could also use them for time-keeping at night. They knew that the skies seem to wheel around us, and so the decans would rise at regular intervals during the hours of darkness.

The Egyptian astronomers couldn't see the fainter decans in the glow of twilight at dusk and before dawn, so typically they'd count a dozen decans rising during the night. That led them to the idea of a 12-hour night. Later, this was matched with 12 hours of daytime – the origin of the 24-hour day that we've inherited today.

But the astronomers of Egypt weren't watching the stars just to while away the hours. The rising of each decan star at night was a matter of life and death. Every passing hour meant that the Sun-god Ra was facing another challenge. Between sunset and sunrise the next day, Ra had to travel underground from the west to the east; and, on the way, there were 12 gates to pass through. If he didn't make it, the following day would never dawn.

'You needed a knowledge of the star-patterns so that the Sun could get through the 12 gates of night at the appropriate time,' explains Chapman. 'They had to have the right prayers, incantations and sacrifices done in the temples at the right time, when you know that Ra is going to be challenged by whichever monster is at each gate.'

As long as they helped Ra through the night, the Egyptians were confident that the next day would dawn; and after centuries of experience, they had every reason to expect that Sirius would bring the life-giving Nile floods every year. But further to the east, in what's now Iraq, the Babylonians were having a harder time of it.

'In Egypt, where the Nile flood came with great regularity, there was essentially no serious development of astronomy,' says Owen Gingerich. 'In Babylon, where you had the Tigris and Euphrates as very capricious rivers, the whole business of omens and astrology developed.' 'This creates a different kind of religion,' Allan Chapman carries on, 'where your gods and your goddesses are often pretty brutal devils who will get you if you get it wrong.'

The Babylonians turned to all kinds of divination to help them propitiate the gods in the right way. They studied flights of birds, patterns of oil poured on water, and especially the state of sheep's livers. But what's most interesting to us is when they turned to the sky.

'They hoped to take something from the regularity of the heavens in order to understand the irregularity of what was below,' Gingerich explains. 'And so mathematical

ABOVE *Sirius, the most brilliant star in the sky, governed the Egyptians' year. They called the star Sothis, and personified it as the goddess Sopdet. When Egyptians first saw Sothis rising just before the Sun, each July, they knew that the Nile was about swell into its annual life-giving flood – caused, they believed, by Sopdet weeping.*

RIGHT *This Babylonian tablet describes observations of Halley's Comet made in September 164 BC. From the seventh century BC onwards, the Babylonians made daily records of the Moon and planets, as well as more exotic celestial events such as comets and eclipses.*

OPPOSITE *A total eclipse of the Sun is the most awe-inspiring of all astronomical events. Ancient descriptions of the Sun disappearing, and the stars coming out, are pretty unambiguous evidence of past eclipses. By analyzing exactly where these eclipses were visible, astronomers can work out how much the Earth's rotation is slowing down.*

astronomy began to develop, very much hand-in-hand with the idea of making astrological predictions. In that sense – before the time of trying to do personal horoscopes – astrology was very akin to a science.'

As Gingerich hints, this astrology wasn't about to tell a Babylonian how her love life was going to shape up next month. As in China, skywatchers would deliver a message of national importance. According to Allan Chapman: 'It was, as far as we can tell, state astrology. In the dodgy world of Mesopotamia, astrology would basically give the King a tip-off when something nasty was going to happen.'

'In antiquity, I wouldn't even distinguish between astronomy and astrology,' says Ed Krupp. 'You see them reading meaning into the sky, so that everything that happens there is believed to have a meaning.'

The Babylonians looked for the significance of the moving Sun and the changing Moon, and the fearsome sight of an eclipse. For instance, on April 15, 136 BC, a scribe in Babylon wrote: 'Solar eclipse: Venus, Mercury and the 'normal stars' were visible; Jupiter and Mars, which were in their period of disappearance, became visible.'

Precision observations like these – and eclipse records from China – have allowed Richard Stephenson to make some astounding deductions about our planet Earth. 'These eclipses are the most significant type of observation to tell us about the Earth's rotation in the past,' he says.

While the Earth spins once a day, it's gradually slowing down all the time. That's why, every couple of years, we have to add a 'leap second' to our clocks. The culprit is the gravity of the Moon, which acts as a brake on our freewheeling planet.

When Stephenson checked out the eclipse of 136 BC, though, he found it should have been visible somewhere to the east of Babylon, not in the city itself. From this – and other eclipses recorded in detail in Mesopotamia and China – Stephenson has had to conclude it's not just the Moon's gravity-brake that is changing the way the Earth spins. Something else must be at work too.

'A lot of it can be attributed to the rise of the land that was glaciated during the last Ice Age,' Stephenson surmises. As the ice sheets melted, the Earth's surface in places like Scandinavia has bounced back: 'Especially in the area round the Gulf of Bothnia, the land's still rising at quite a rate.'

The Babylonian astronomers were unaware of the scientific insight they would bring to the study of our planet. In their search for some underlying order in the Cosmos, their sights were firmly set on the motions of the other planets in the sky.

The earliest account from Babylon describes the planet Venus, around 1700 BC, as retold in a much later record. After the Assyrians had conquered Babylon and left again, astronomy picked up once more in the eighth century BC.

The evidence comes in literally thousands of fragments of clay tablets. Marked with a stylus in distinctive cuneiform script, they record sightings of the Moon and planets, and attempts to predict eclipses. And to identify how the planets moved, the Babylonians had to group the background stars together into constellations. Some of these star patterns have survived right though to the present day.

'I think the story of the constellations hasn't been told, and it may never be told fully,' says Ed Krupp. Some of our most striking constellations – Orion the hunter, Gemini the twins and so on – have come down to us from Mesopotamia. But Krupp

BELOW *The Great Bear is perhaps the most ancient of all constellations, found in Europe, Asia and even among the native peoples of North America. In this illustration from Johann Bode's great star atlas of 1801, the seven stars of the Plough (Big Dipper) can be seen in the bear's back and tail.*

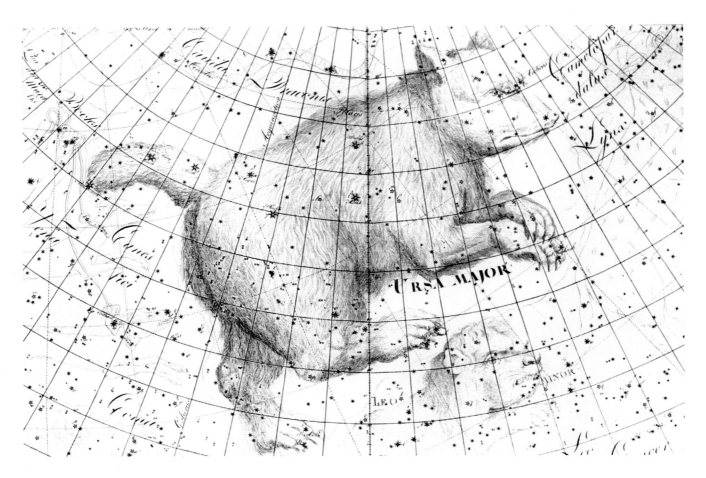

contends that there wasn't a single point in time when people sat down and divvied up the sky for the fun of it. 'One of the great myths in astronomy books,' Krupp maintains, 'is the very romantic notions we have about our ancestors' response to the sky – that the shepherds at night had nothing better to do, and so began telling stories about the stars to entertain their friends.'

Among the oldest constellations may be the seven stars of the Plough or Big Dipper, which is part of a star group the Greeks called the Great Bear. Amazingly, some of the Native North American legends also see a celestial bear among these stars.

Brad Schaefer – a professional astronomer and historian of skywatching – thinks that the Great Bear in the sky is the symbol of a bear cult that stretched all around the Arctic long ago, and was carried into North America with the first human migrations from Siberia. Perhaps it dates back to the time when our Palaeolithic ancestors erected shrines to cave bears in Europe, some 30,000 years ago.

But Ed Krupp remains to be convinced. 'When you look into details of these arguments, they're forged out of a piece of information here, and here, and they're not really testable – they really don't hold up well.'

Krupp is willing to believe, though, that the Great Bear constellation came to Mesopotamia from Asia. He thinks the Pleiades – the Seven Sisters – may have followed a similar route, because you find myths about a 'missing' star in this cluster everywhere from central Asia to India to Babylon.

'But the oldest place we can go where we've got documentation,' Krupp emphasizes, 'is Mesopotamia. Here we find the constellations of the Zodiac. The bull (Taurus) is very clearly present at that point. On other hand, the Aries – the ram – is not – the Mesopotamian version of Aries was 'the date labourer.'

The Babylonians focused on the constellations of the Zodiac because they were essential to their astrology. These stars provided the path that was followed by the wandering Moon and planets.

And the Babylonians gave characters to the planets. Brilliant Venus was Ishtar, the goddess of love; while reddish Mars was Nergal, the god of war. They saw stately Jupiter as Marduk, the god of Babylon and hence the head of their pantheon.

These are the three most prominent worlds in the night sky, and we can see how the Babylonian names have come down to us – through the Greeks – in the Latin versions of the same gods and goddesses.

The people of Babylon saw Saturn – our god of old age - as the hero Ninurta. And the swift messenger of the gods, Mercury, appeared in the skies of Mesopotamia as Nabu, the god of wisdom and writing.

Along with their meticulous observations of the sky, the Babylonians developed some pretty clever mathematics. Instead of being based on 10, though, their number system hinged around 60. That's why – to this day – we have 60 minutes in an hour, and 360 (6x60) degrees around a circle.

That made the Babylonian astronomers the first people who could work out where the planets would lie in the future. Each year, they'd draw up a guide as to what would

ABOVE *Orion is one of the most distinctive star-patterns, delineating the shape of a giant in the sky, with three stars marking his belt and a cluster of stars – plus the luminous Orion Nebula – forming his sword. According to ancient Babylonian legends, this sky-figure represents the great warrior Gilgamesh, who lived around 2500 BC.*

be happening in the sky for the coming months: these Goal Year Texts were like the astronomical yearbooks you can buy today, but with all the astrology added as well.

For instance, one cuneiform tablet records: 'Nabu (Mercury) is visible with Nergal (Mars) at sunset; there will be rains and floods. When Marduk (Jupiter) appears at the beginning of the year, in that year the crops will prosper.'

The Babylonians were certainly obsessed with the sky. But they weren't interested in the structure of the heavens; they weren't bothered to understand why the planets move the way they do. They just did the sums and hoped they'd get it right enough to keep the gods happy.

And there's one prediction from Babylon that has kept more people arguing – to this day – than any other. 'Where is he that is born King of the Jews? For we have seen his star in the East, and we have come to worship him.'

The 'Wise Men from the East,' described in St Matthew's gospel, were probably astrologers from Mesopotamia, steeped in the lore of Babylonian astronomy. But what had they seen – and when? The second question isn't as trivial as it sounds. A sixth century monk, Dionysius Exiguus, made a mistake in calculating the date of when Jesus was born. So the virgin birth was not in AD 1, but actually several years BC.

What we do know is that it wasn't the brilliant star we see on Christmas cards, outshining even the Moon in the sky. The Chinese would certainly have added such an amazing apparition to their list of 'guest stars'.

We put the question to our team of eminent astronomical historians. To our surprise, the answers are pretty diverse.

Ed Krupp throws the question back at us: 'You give me a date for the birth of Christ, and I'll give you a Star of Bethlehem.'

'7 BC' we reply. That's based on one of the earliest ideas about the Star, put forward by the great German astronomer Johannes Kepler – he who discovered the way the planets orbit the Sun. In 1604, Kepler had been watching the planets Jupiter, Mars and Saturn clump together in the sky. As they did so, a brilliant star blazed out nearby. We know today that it was a distant – and totally unrelated – supernova. But Kepler drew the obvious conclusion: that the close conjunction of the planets had by some astrological power kindled the great star.

Kepler calculated there was a similar grouping of planets in 7-6 BC. At that time, he surmised, they would have ignited a similar brilliant celestial orb: the Star that invited the Magi to Palestine. This planetary grouping – with or without the impossibly bright star that Kepler invoked – must have given off strong astrological vibes. God the Father (Saturn) made three separate approaches to the King of the planets (Jupiter) in the astrological sign of the Jews (Pisces). The omens were clearly written in the sky: the Messiah was about to be born in Palestine.

LEFT *The Magi attending the infant Jesus were astrologers who inherited the lore of ancient Babylon. Detail of* Adoration of the Shepherds *by Pinturicchio.*

'Well in 7 BC, certainly, we can apply the multiple conjunctions of Jupiter and Saturn,' Krupp comes back, 'but in fact this attempt to recreate the Star of Bethlehem from the available data is a romance. There's hardly any data – just a couple of references in the Book of Matthew, none of them terribly specific.'

So, we turn to David Hughes, who several years ago wrote a book called The Star of Bethlehem Mystery. 'You have this sort of change-over of power from one regal planet to another,' he expounds, 'and that would have at least have got a few Magi off their backsides and onto camels and traveling to Jerusalem.'

But even Hughes is not entirely convinced. 'At the back of my mind, there is the very strong thought that dear old Matthew was a spin-doctor of the day, and he made the whole thing up. He thought: I've got to promote this new Jesus, get the Jews Christianised; what do I need to really emphasise the Messiah nature of this man – I know, a bright star!'

On the other hand, Owen Gingerich at Harvard believes we can take the Bible at its word, provided we get into the minds of the Magi. He cites the work of the scholar Michael Molnar, who was convinced that the astrologers wouldn't be excited by what they saw in the sky, but what was shown on their charts.

'He looked at the few known horoscopes for Roman Emperors and seeing what seems to be required for royalty,' Gingerich explains. Also, Molnar found coins from the time that show that Palestine was represented by the sign Aries (rather than Pisces).

'And this could help you understand why the Magi arrived in Jerusalem,' Gingerich elaborates, 'saying there's this striking configuration that is representative of

BELOW *Many astronomers think this is how the Star of Bethlehem really looked – an astrologically important conjunction of bright planets.*

a royal birth. And if the court was not particularly into astrology, they could then be taken by surprise and astonishment.'

The 'Star,' according to this theory, was an astrological event that took place during the daytime, so no-one could actually see it: on April 17, 6 BC, the Moon passed right in front of Jupiter, in the skies above Jerusalem. At the time, all the planets formed a tight cluster around the Sun – in the Jewish constellation of Aries. But Allan Chapman at Oxford disagrees with even the best astrological account. 'What that thing was, we don't know,' he emphasizes. 'As far as I'm concerned, it was a miraculous event.'

What we can probably all agree is that there'll never be an answer that convinces everybody. And it's one of those wonderful evergreen stories that will never go away, resounding as it does from the beginning of one of the world's great religions.

'What's important is not whether there really was a particular object,' Ed Krupp declares, 'but the fact that people are always assigning significance to what happens in the sky. And when they want things on Earth to be significant, they make sure that something happened in the sky.'

Wheels within wheels

Greek diver Elias Stadiatos was in for the surprise of his life that day in October 1900. As he dived 200 feet (60 m) deep for sponges, off the coast of the island of Antikythera, he spotted what was clearly a shipwreck. Swimming closer, he saw intriguing shapes…and then panic struck.

'There's a heap of naked women down there,' Stadiatos babbled to his crewmates as he surfaced. He described rotting corpses of people and horses on the sea bed.

Fearing his colleague was suffering hallucinations from the compressed gas he was breathing, the ship's captain dived down – and surfaced with the arm of a bronze statue. Over the next two years, archaeologists recovered a rich trove of bronze and marble statues, wine and jewelry – and one highly corroded lump of metal, about the size of shoe-box, that appeared to have gears embedded in it.

Since its discovery, this unique artifact from the seabed – the Antikythera Mechanism – has amazed scientists almost as much as the statues perturbed Elias Stadiatos.

It has turned out to be the first computer: a clockwork mechanism, a millennium older than any other known, which was used to predict precisely the positions of the Sun and the Moon in the sky – along with eclipses, the most momentous of celestial sights.

ABOVE & RIGHT *As seen in this reconstruction, the front dial of the Antikythera Mechanism indicated the position of the Sun and the Moon against both the Greek Zodia and the Egyptian calendar. Within the world's oldest computer, an intricate system of gear-wheels worked out the phases of the Moon, the occurrence of eclipses, and perhaps the motions of the planets.*

The circular gear wheels of the Antikythera Mechanism reflect the ancient Greeks' preoccupation with circles – and with the idea that everything in the sky moves around in circular paths, because the heavens are the home of perfection, and a circle is the ideal shape.

While the Babylonians and the Egyptians believed that gods and goddesses were always busy pushing the Sun, Moon and planets along, the Greeks took an extraordinary step. They removed the supernatural beings from the frame. Instead, the celestial bodies moved because of something in their own nature. And human beings could work that out.

In short, the ancient Greeks invented science.

The world's first scientist was Thales of Miletus. This pioneering Greek thinker lived not in Greece itself, but in one of the Greek colonies near what's now the busy tourist resort of Bodrum in the southwest of Turkey.

Today, Thales's home town is nothing but an archaeological site. Around 600 BC, it was the most powerful city of the region, boasting four harbours where merchant fleets plied their trade to a dozen cities under Miletus's control. But its river, the Meander, lived up to the reputation which would give its name to wandering watercourses worldwide. The Meander strayed from its original bed, the harbours of Miletus filled with silt, and the town eventually died.

But in Thales's time, Miletus was buzzing with ideas. The astronomy of the Babylonians filtered in from the east; while Thales was able to sail to Egypt and absorb their knowledge – in particular, the geometry that the Egyptians used to stake their own claims after the Nile flooded each year and covered the land with fresh life-giving mud.

Thales was inspired by their ideas, but not their superstitions. He was the first person to suggest that natural forces are responsible for events in the world around us. Earthquakes, for instance. Until then, people thought that earthquakes were the sign of an angry god shaking the ground. But Thales suggested a 'scientific' explanation: that the Earth floats on water, and earthquakes are caused by giant waves passing underneath.

Thales thought hard about the sky, too – so hard, in fact, that he was once so busy looking heavenwards that he fell into a ditch. An old woman answered his cries for help. She pithily asked how he could hope to know anything about the stars, when he didn't even know about the Earth beneath his feet!

Drawing on the Babylonians' astronomy, Thales apparently made the first accurate prediction of a total eclipse of the Sun, on May 28, 585 BC. At the time, two of the nations in the region – the Medes and the Lydians – were in the midst of a pitched battle, at the climax of a fifteen-year war for control of present-day Turkey.

The Greek historian Herodotus recorded: 'a battle took place in which it happened, when the fight had begun, that suddenly the day became night. And this change of the day Thales the Milesian had foretold. The Lydians however and the Medes, when they saw that it had become night instead of day, ceased from their fighting and were much more eager both of them that peace should be made between them.'

Thales wasn't just an early nerd. After making the first scientific prediction of a bumper crop of olives, he bought up options on all the presses in the region and made a small fortune by cornering the olive oil market.

Following Thales, science blossomed along the coast and on the islands within a day's sail of Miletus. This Scientific Revolution was an incredible leap for humankind, and it was concentrated within a remarkably small part of the globe. In some ways, it

BELOW *A roaring lion flanks the entrance to the great amphitheatre in Miletus, now in Turkey, which – over two millennia ago – saw audiences of 15,000 people flocking to cultural and sporting events in one of the most important cities of the ancient world.*

was like the Industrial Revolution that would later bloom in the Midlands of England, and the recent Information Revolution in California's Silicon Valley. But why?

'I think the reason,' opines Allan Chapman, historian of astronomy at Oxford University, 'is to do with the fact that the Greek world was utterly fragmented geographically.' The Egyptians were based on the great river Nile, while the ancient Mesopotamian civilizations were founded in the valleys of the Tigris and Euphrates. 'These were big river cultures,' Chapman continues, 'with excellent control systems. The river could be a source of authority for armies, for moving goods and keeping people under control.' Greece, on the other hand, was broken up into innumerable valleys and islands. There could be no over-arching control system. Instead, each region developed its small own city state, or polis – the Greek word that gives rise to 'politics'. And trade between the city states was in the hands of independent merchants.

'So Greece developed what I call the world's first middle-class culture – merchants, traders, trade your own money, keep your own money, don't get it scythed

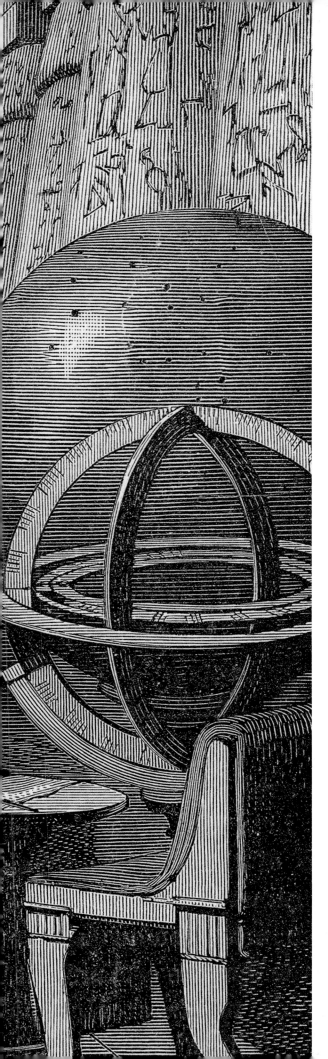

LEFT *From Alexandria, in Egypt, a Greek astronomer uses a cross-staff to measure the positions of stars in the constellation of Orion. The astronomer in question is Hipparchus (ca. 190-125 BC), who compiled the world's first accurate star map.*

off,' Chapman enthuses. With fluid funds, the prosperous Greek middle class had time on their hands – and freedom to do what they liked with it.

'When you're sitting around drinking your wine and your olive oil,' says Chapman, 'you want entertaining. You patronize architecture, you patronize theatre, you patronize intellectual life. That's why you have plays by Aristophanes, which are uproariously funny even to us today. And I think that science is simply part of that incredible burgeoning of Greek intellectual life.'

Thales's reputation became a magnet for thinkers from the surrounding region. As he reached old age, he received a visit from the son of a merchant based on the nearby island of Samos. The lad's name was Pythagoras.

On Thales's recommendation, Pythagoras traveled to Egypt to bone up on their ancient wisdom. As well as a good grounding in geometry, he also picked up a lot of their ideas on secret spiritual practices. He was then taken prisoner by the Babylonians. But he put his time in captivity in Babylon to good use. In the words of a later Greek philosopher, Pythagoras 'reached the acme of perfection in arithmetic and music and the other mathematical sciences taught by the Babylonians.'

When Pythagoras returned to Samos, he was unconventional even for a philosopher. He taught his pupils in a cave outside the town, using weird Egyptian symbols. Eventually, he emigrated to the Greek colony of Croton (now Crotone), on the sole of the 'foot' of Italy. Here, Pythagoras founded a secret school, where all his followers had to be vegetarian and relinquish their personal possessions. Unusually for Greece, where men and male friendships were paramount, Pythagoras's school was open to both sexes, and many leading members were women.

Today, we learn about Pythagoras at school by his theorem on right-angled triangles. That was just one minor piece of maths he'd picked up on his travels. What drove Pythagoras was the idea of harmony – harmony in maths, harmony in music and harmony in the heavens.

Apparently, the great philosopher was listening to a blacksmith striking his anvil with various hammers one day, when he noticed something intriguing: a smaller hammer always rang with a higher-pitched sound.

Pythagoras was a keen musician, playing his lyre to help cure the sick. Inspired by the harmonious blacksmith, he now experimented with strings of different lengths, and discovered that he could generate the notes of the musical scale with strings whose lengths were in simple ratios to each other.

He believed there must be a similar mathematical order in the heavens above. Ever the master geometer, Pythagoras taught that the movements of the Sun, Moon and planets must be based on the simplest and purest geometrical shape, the circle.

And Pythagoras's obsession with circles led him to the first major discovery in astronomy. He realized that the Earth is not flat, but is a round ball floating in space.

Today, we take this for granted. But it's not all obvious – and none of the other great civilizations of the past took this giant intellectual leap. After all, if the world is round, why don't you fall off?

Pythagoras based his revolutionary idea on eclipses of the Moon. He noticed that the edge of the Earth's shadow falling on the Moon is always curved. The only

ABOVE *One of Pythagoras' master-strokes was to realise that the Earth is a sphere. Observing a lunar eclipse, he noticed that the shadow of Earth cast onto the Moon is curved. In this recent image of a lunar eclipse – taken in Rio de Janeiro on May 15, 2003 – the shadow is clearly visible on the right. The curvature also reveals that our planet is about four times the diameter of the Moon.*

RIGHT *Pythagoras tests the notes emitted by a range of water vessels against a series of numbered bells, struck by his assistant. The relationship between music, number and harmony was crucial to many of the Greek philosophers. This woodcut dates from* Theorica Musices *– a book published in 1492 by Gaffurius of Milan, then one of Italy's leading music experts.*

PYTAGORA

LEFT *A Roman mosaic of Plato's School of Philosophy, based in Akademeia – a northern suburb of Athens. It is to Plato that we owe the word 'academia' – and the concept of a centre for learning. The School boasted a gymnasium and a centre for worship, just like universities today. Closed down by Emperor Justinian in AD 526, the site was extensively excavated by archaeologists in the twentieth century.*

way the Earth can cast a circular shadow, whatever the direction of the Sun, is if our planet is a sphere.

Thanks to the adventures of his early years, Pythagoras also knew he could see new stars as he traveled south. This didn't make sense if the Earth was flat; but would naturally be the case if the world was curved, and as you travel farther south you are looking round the Earth's curve to a region of the sky always hidden from Greece.

From 500 BC or so, all educated people knew that the Earth was round. And so it's a hoary old myth that Columbus was worried that he might he might fall over the edge, as he headed out into the Atlantic in AD 1492.

Around the spherical Earth, said Pythagoras, the Sun, Moon and planets all sailed in circular paths. Like the strings of his lyre, these invisible circles hummed as the planets moved, emitting the deep and melodious Music of the Spheres.

Well after Pythagoras's death, a soldier and aspiring politician visited Croton. His name was Plato. He was bowled over by the beauty of the mathematics at Pythagoras's school, and it shaped his view of the world. What we see around is constantly changing and deceptive, Plato taught. The only thing that is real is a world of ideas, including pure mathematics.

In 387 BC, Plato set up a school of philosophy in Athens, the Academy. Pythagoras would have been proud of the inscription over the door: 'Let no-one ignorant of geometry enter here.'

Plato had no doubt that the hidden motion behind the Sun, Moon and planets must be in the form of circles. But there was one catch. While the Sun and Moon move around the heavens at a roughly constant speed, the planets are a much more

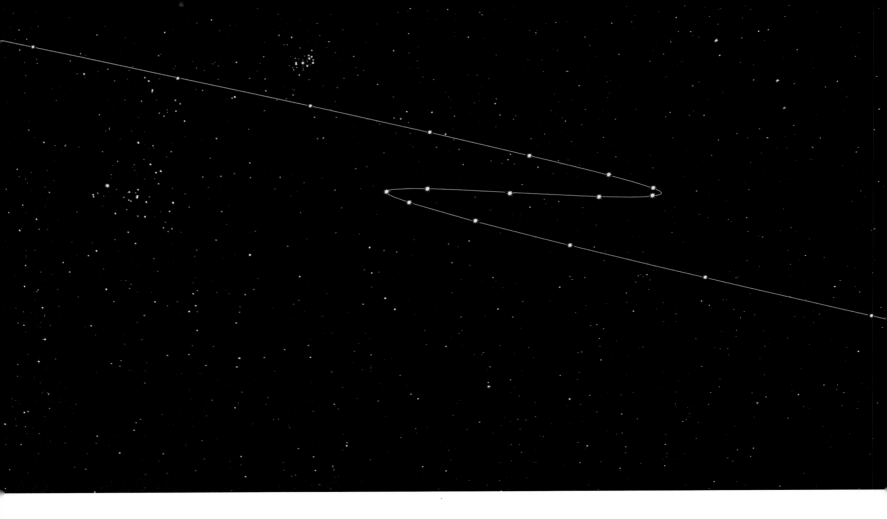

ABOVE *Time-lapse image of the planet Mars – taken over several weeks – looping under the Pleiades star cluster. The Greeks came up with elaborate explanations for these 'retrograde loops', assuming the planets circled the Earth. Now we know that the loop is caused by Earth overtaking Mars in its own orbit around the Sun – making Mars appear to slip backwards.*

unruly bunch. Mars, for instance, makes huge loops, backwards and forwards through the constellations.

The challenge of the planets' loopy motions was taken up by a follower of Plato, Eudoxus, who was based in the great crucible of scientific thought in southwestern Turkey. Born in Cnidus (now Tekir) around 410 BC, Eudoxus proposed that the Earth is surrounded by a set of mathematical spheres, like the layers of an onion. Each planet is fixed to a sphere, and is carried around as that sphere rotates. Eudoxus cleverly managed to mimic the planets' complicated paths by giving each planet four spheres that are linked at their poles, and turn at different rates.

It was a nice try; and everyone was pretty pleased at the time. The only problem was that his rotating spheres wouldn't have fitted the motions all that well. 'It fails completely for Mars,' comments Harvard historian Owen Gingerich, 'but 'so what' we say – it's a theory to be admired, not to be used.'

Plato's successor in Athens, Aristotle, certainly admired Eudoxus's spheres. Aristotle was very much the opposite of Plato. He wasn't one for abstract maths, but believed firmly in what we actually see around us. Everything we meet in everyday life, he said, is made up from four 'elements': earth, water, air and fire. These form the changing world around us.

But the heavens are entirely different. The Sun, Moon, planets and stars – along with the spheres that carry them – are perfect. They are made of a pure substance we don't find on Earth: quintessence.

'This 'fifth' element that was weightless, transparent and so on was a serious error which impeded the later development of science,' says Oxford's Allan Chapman. 'It created such a great dichotomy between the heavens and the Earth. But on the other hand I can't really fault Aristotle for it, in the sense that the heavens do appear to be fundamentally different from our mundane surroundings.'

The pupil of Aristotle's who would pave the way for the next breakthrough in Greek astronomy wasn't a scientist; or even a scholar. He was the world's greatest ever military commander: Alexander the Great. In his short lifetime – he died before he was 33 – Alexander conquered the known world, from Greece to India. On the way, he destroyed the highly civilised Persian Empire, which had inherited the astronomy of the Babylonians, replete with detailed predictions based on a load of complex arithmetic.

This new influx of knowledge gave the Greeks a major wake-up call. Their beautiful circular motions in the sky weren't enough. They would have to actually measure what was going on in the heavens – and as accurately as they could.

On Pythagoras's home island of Samos, an astronomer named Aristarchus accepted the challenge. 'Aristarchus is very interesting because he was a mathematical astronomer,' explains Owen Gingerich. 'He's wanting to put numbers on things; he's not just doing vague philosophical speculations.'

Aristarchus started by applying his mathematical genius to the Sun and Moon. By watching carefully when the Moon was swallowed up by the Earth's shadow during

BELOW *The magnificent Parthenon was new when Plato set up his Academy in Athens. While this temple celebrated the goddess of wisdom, Athene, Plato pursued wisdom in the abstract.*

ABOVE *Down-to-Earth Aristotle
– as seen in this formidable bust on show
at the National Museum of Terme,
Italy – was the complete opposite of his
predecessor, Plato. He believed in a
division between the 'elemental' Earth
and the 'pure' celestial firmament.
His ideas held sway for 2000 years.*

a total lunar eclipse, Aristarchus deduced that the Moon is half the diameter of our world. (In fact, the true answer is more like a quarter, but it wasn't too bad for a first attempt).

Later, Aristarchus watched the Moon waxing in the sky until was exactly half-lit by the Sun. He measured the distance between the Moon and the Sun at that moment. Using simple geometry, he could then work out how much farther the Sun lies from us, as compared to the Moon. The answer was 19.

Aristarchus also knew one of the great coincidences in the sky: that the Sun and Moon appear to be the same size. So, if the Sun lies 19 times further away, it must be 19 times wider than the Moon.

And that was astonishing. That meant the Sun must be almost ten times larger than the Earth. Actually, his primitive measurements severely underestimated the Sun's size and distance: today we know that the Sun is a hundred times wider than our planet. But even ten times bigger seemed to reverse the importance of the Sun and the Earth. If the Sun was so much bigger, why was it circling the Earth? Aristarchus came up with a revolutionary new idea. The Earth was actually circling around the Sun.

No-one accepted this idea for almost 2000 years, when the Polish canon Nicolaus Copernicus came up with a whole raft of good reasons why the Sun is stationary, and our world is just a planet traveling around it once a year.

So why did the Greeks not make this great leap, when they had it in their grasp? First, bigger wasn't necessarily better: just because the Sun was bigger than the Earth didn't make it any more massive. After all, everything in the sky was made of quintessence, which was naturally weightless.

Eclipſis   Soliſ

'And also,' says Allan Chapman, 'because these early Greeks were remarkably playful we don't know how much Aristarchus believed it himself.' We pick out Aristarchus because his idea chimes in with what we know today; but it was just one of the concepts that these early cosmologists were playing with. The followers of Pythagoras, for instance, thought the Earth and Sun both move, as they revolve around a central fire.

In fact, the only surviving reference to the moving-Earth theory comes in a sentence or two written by another Greek scientist, Archimedes: 'Aristarchus has brought out a book consisting of certain hypotheses – that the fixed stars and the Sun remain unmoved, that the Earth revolves about the Sun on the circumference of a circle…'

'And Archimedes was trying to do something very different,' says Gingerich, 'namely to write a number large enough to be larger than any number you would need.' And he did this by imagining how many grains of sand you'd need to fill the Universe.

'I suspect that he and Aristarchus were sitting under an umbrella in the market place,' Gingerich continues, 'arguing about this, and just to be feisty Aristarchus says 'yeah, but suppose the Earth is not fixed but the Sun is, then in order not see any effects of this motion, you've got to make the Universe a thousand times bigger in volume.'

Faced by this hypothetical challenge, Archimedes simply cranked up his number-system to prove it could do anything. Gingerich suspects Aristarchus never wrote up his idea: 'I bet it was just something written on the tablecloth as theoreticians are wont to do.'

Archimedes lived in the Greek colony of Syracuse, in Italy. He had a huge flair not only for maths, but also for practical science and engineering. As every schoolchild knows, he discovered the rules for how objects float while he was in the bath-tub, and rushed down the street naked shouting 'Eureka!' ('I have found it!').

ABOVE *Aristarchus was notorious for calculating that the Earth circles the Sun, rather than vice-versa. In this woodcut, a smiling Moon eclipses a disgruntled Sun, causing darkness to fall on the Earth. Because of solar eclipses, Aristarchus knew that the Moon and Sun appear the same size in the sky: from his observations of the Moon he could then work out that the Sun was ten times larger than the Earth, and so was the natural centre of Universe.*

LEFT *Archimedes – ever-fascinated by numbers – has his Eureka-moment. He finds that an object is buoyed up in water with a force equal to the weight of the liquid it displaces. It's not known how his neighbours reacted to his naked canter down the street afterwards, but his discovery has implications even today in our ocean-dominated world.*

His inventions also helped to save Syracuse from a Roman invasion. When the city was eventually over-run, one legend says a soldier killed Archimedes out of frustration, when the great mathematician refused to move away from his latest mathematical proof, muttering 'Don't disturb my circles.'

Even more interesting, the Roman writer Cicero declared that the spoils from Syracuse included an interesting astronomical device. 'Archimedes fastened on a globe the movements of Moon, Sun and five wandering stars [which] made one revolution of the sphere control several movements utterly unlike in slowness and speed.'

Think of the 'globe' as a circular dial, and this is an almost exact description of the later Antikythera Mechanism. It suggests Archimedes was the godfather of the world's first astronomical computer: with his unique combination of mathematical and engineering genius, he certainly had all the right qualifications.

The actual Antikythera Mechanism didn't come from Syracuse, though. It was shipwrecked on a vessel that was heading towards Rome from the Aegean Sea. The vases and other treasures on the ship point the finger at Rhodes.

Today, mention of Rhodes may bring to mind wild holiday nights spent at discos – or the massive fortifications built by the Knights of St John (the Knights Hospitallers) in the Middle Ages. But Rhodes had a key place in history well before then.

The island lies just off the Turkish coast, near Samos and Miletus, in the homeland of the Scientific Revolution. In 305 BC, Rhodes town withstood a major siege, and to celebrate the inhabitants built a huge bronze statue of Helios. Standing two-thirds as tall as the Statue of Liberty, it justly entered the roll-call of ancient Wonders of the World as the Colossus of Rhodes.

A hundred and fifty years later, Rhodes was to be home to a colossus of the astronomical world. His name was Hipparchus; and for all his academic fame, he remains a shadowy figure. Hipparchus was born in what's now Turkey, and died in Rhodes, but as a person we know very little about him. What's undoubted are his astronomical achievements. Hipparchus was ideally placed to firmly fuse Greek astronomy with the heritage of observations from Babylon.

His list of 'firsts' may seem rather dry, but they form a quantum leap in humankind's knowledge of how the heavens work. He measured the length of the year to amazing precision – better than 7 minutes in 365 days. It was so accurate that he came across a puzzling contradiction.

If you measure the year by the way the Sun moves against the background of stars, it's 20 minutes longer than the year you measure from Midsummer to Midsummer. Hipparchus called this effect the Precession of the Equinoxes – and it has lot of consequences for astronomy. The stars seem to swing around with the same motion, completing one circuit of the heavens in 26,000 years, so the star that Hipparchus saw marking the north pole of the sky is not the same Pole Star we observe today. And the constellation where the Sun crosses the equator of the sky has changed from Aries in Hipparchus's day to Pisces, and now – as hippies of the 1960s will remember – we are entering the Age of Aquarius.

Hipparchus spent his nights carefully measuring up the 850 brightest stars on the sky, and inscribing their precise positions on a globe. Though his original sphere is long since lost, American astrophysicist and historian Brad Schaefer claims that you can see a copy of this ancient star map today.

The Farnese Atlas is a marble statue in Naples, depicting the giant Atlas holding up a globe of the stars. Relief figures on the marble ball show 41 constellations from ancient times. The Farnese globe is Roman, but it's undoubtedly copied from a Greek original.

Schaefer has studied the positions of the constellations carefully. They will change with time, because of the Precession of the Equinoxes. And he's concluded that the original globe must date from around 125 BC. And the only astronomer who was mapping the stars at that time was Hipparchus.

Historian Ed Krupp from the Griffith Observatory in Los Angeles calls Schaeffer's discovery 'a clever and disciplined analysis that reveals unexpected roots of scientific astronomy in a celebrated work of ancient art.'

Hipparchus also checked out the Moon's motion in extraordinary detail, which meant he could predict eclipses with more confidence than anyone before him.

Which brings us to the Antikythera Mechanism. In 2006, an international team from Greece and the UK probed its innards in unprecedented detail. They found 30 gear wheels inside, all hand-cut from bronze plate, the largest bristling with 235 teeth. An ancient astronomer could operate it by turning a handle on the side.

According to their report in the journal Nature, the team found a pointer which represents how the Sun moves against the signs of the Zodiac, and another showing the Moon's motions. A small ball, silvered on one side, rotated to reveal the phases of the Moon. The complicated gearing even mimics the way the Moon's speed changes over a cycle that lasts nine years.

Mike Edmunds, one of the team members from the University of Cardiff, says 'The way in which it displayed the Moon is very advanced, and seems to be based on a theory of the Moon developed by Hipparchus.'

So, was the ultimate Greek geek behind the world's first computer? 'It's very tempting to think so,' concurs Edmunds. 'We haven't actually found his fingerprints or 'Hipparchus made this mechanism' but whoever did build this was extremely intelligent.'

BELOW *Rhodes: a crucible of history, and home to Hipparchus. Over the centuries, myriad cultures have stamped their intellectual and archaeological identities on the small island.*

The team also managed to decipher inscriptions on the bronze back of the Antikythera Mechanism. They describe the cycle of eclipses of the Moon and Sun. An eclipse of the Moon is interesting but not that unusual. To see a total eclipse of the Sun, on the other hand, is the most stunning of all sights in the sky. The Antikythera Mechanism is probably the first computer that could predict a total solar eclipse.

And it was an eclipse of the Sun in March 2006 – predicted this time by NASA's computers to occur on the borders of Egypt and Libya – that brought us to the city that saw the final flowering of Greek astronomy. Alexandria was founded by Alexander the Great when he conquered Egypt. His general Ptolemy brought Alexander's body back here after his death, and declared himself King Ptolemy I. The city was the country's leading Mediterranean port, at one time the largest conurbation in the world. And it boasted another of the Seven Wonders of the World, the great lighthouse Pharos.

Today, we find Alexandria a city of contrasts. The back streets bustle with shops and Arabic market stalls, while the long sea-front is lined with gardens and graceful European-style buildings. The highlight of any visit is a vast glass building at one end of the harbour, shaped like a giant disc tilting up from the ground. Inside, you can ascend 11 staggered levels, past exhibition halls, conference rooms – and endless shelves that will eventually carry 8 million books.

This is the Bibliotheca Alexandrina, and it's the spiritual successor to the ancient Library of Alexandria – the repository of all the knowledge of the ancient world. And the most coveted academic position was to be the Librarian.

The third Librarian of Alexandria was a Libyan polymath called Eratosthenes. Appointed in 240 BC, he was reputed to be overbearing and generally disliked: his

ABOVE *A Roman statue of Atlas carries a copy of Hipparchus's Farnese globe – his incredibly accurate star catalogue – on his shoulders. Brad Schaefer, an astrophysicist and historian holidaying in Italy, stumbled across the globe in the National Archaeological Museum in Naples. He has dated the original to around 125 BC – contemporary with the time when Hipparchus lived.*

ABOVE *Repository of all knowledge: the great Library of Alexandria, where the scientist Eratosthenes held the coveted post of Librarian. In this later engraving of the Library in its heyday, scholars absorb themselves in examining scrolls.*

colleagues nicknamed him Beta (the second Greek letter) on the grounds he wasn't first in any field of knowledge. But Eratosthenes was an Alpha scientist in one respect: he made the first accurate measurement of the size of the Earth.

As he went through the hundreds of thousands of scrolls in the Library, Eratosthenes learnt a strange fact about the town of Syene (now Aswan), far to the south at the First Cataract of the Nile. On Midsummer's Day, the Sun passed directly overhead – so that it shone vertically down the wells in the town.

At the next summer solstice, Eratosthenes carefully checked the Sun's height as seen from Alexandria. It certainly lay some way off the overhead point. This was because Alexandria lay around the curve of the Earth from Syene, Eratosthenes deduced. The royal surveyors provided him with the actual distance from Alexandria to Syene, and with a little geometry Eratosthenes worked out that the circumference of our globe is 252,000 stadia.

Greeks measured their distances in stadia, the length of a sports stadium. The problem is that we don't know exactly how long that was. But the best guess would make the Earth between 24,670 and 28,950 miles (39,700 and 46,600 km) round. That's amazingly close to its actual size, 24,880 miles (40,040 km). In fact, Eratosthenes's measurement is better than the value that Christopher Columbus used when he set off across the Atlantic some 1700 years later.

Another leading light of Alexandria was the mathematician Apollonius, from Perga in what's now Asia Minor, who was known as the Great Geometer. He applied his formidable skills to the wayward motion of the planets. Why did Mars, for instance, sometimes change its direction through the sky, make a large backwards loop, and then move forward again?

Apollonius 'knew' that everything in the sky had to move in circles, so he began to play around with circles of different size. And he hit upon an ingenious answer. Mars doesn't simply travel around the Earth in a large circle, as Aristotle had said. It's fixed to the rim of a smaller circle, which is carried around by the larger circle.

Visit a funfair, and you get the idea. On a Waltzer, you sit in a car that spins around, all the time that the car is speeding around the central pivot. Sometimes you're going forwards, sometimes backwards. The operator in the centre sees you traveling erratically, in rather the same way that Mars appears to go around the Earth.

The small circle is called an epicycle; and this theory was to dominate astronomy for almost two millennia. By an enormous irony, though, Apollonius's other great love was a set of curves that included ellipses and parabolas. He painstakingly pioneered the geometry of ellipses for future generations of mathematicians – and we now know that the Moon and planets actually travel in ellipses.

Over the next couple of centuries, Alexandria was taken over by the Romans, and saw Cleopatra dallying first with Julius Caesar and then with Mark Antony. In Palestine, the Romans sacked the holy city of Jerusalem, a few decades after Jesus Christ was born.

Around the year AD 150, Alexandria saw the last major skywatcher of ancient times: Ptolemy. Confusingly, he had the same name as the kings of Egypt, though as far as we know he wasn't related. Later artists hedged their bets by drawing pictures of Ptolemy observing the sky wearing a crown!

'I think Ptolemy was the greatest astronomer of antiquity,' says Owen Gingerich. 'He was an incredible organizer, putting out a huge number of different treatises on essentially every form of mathematical science that he could think of. And he's very ingenious in how he arranges this material.'

Ptolemy drew together all of the Greeks' knowledge of the heavens in a 13-volume masterpiece called the *Mathematike Syntaxis* (*Mathematical Treatise*), better known today by its later Arabic title, the *Almagest*.

The *Almagest* was the 'Bible' of astronomy for 14 centuries – how many authors wish that their books had that longevity! Most of its content, as Ptolemy himself admits, was based on the great scientists who had gone before, especially Aristotle and Hipparchus. But Ptolemy forged them into a single theory that seemed to provide the final answer to the Universe. And he buttressed his great astronomical fortress with some powerful new maths.

The Earth, Ptolemy began, was the centre of the Universe, as Aristotle had told us. Bowing towards Hipparchus, he explained how the Sun and Moon move around the Earth. Next, Ptolemy creates the sky that all amateur astronomers know today. He gives a definitive list of 1022 stars, and how they are arranged into 48 constellations, including the Great Bear, Orion, Leo and Scorpius. Nowadays, we have an additional

BELOW *A phoenix rises from the flames: Bibliotheca Alexandrina is the modern incarnation of the ancient Library. A wonderfully futuristic building, it is home to six specialised libraries, three museums, seven research institutes, exhibition areas and a planetarium.*

Ptolemy's universe, as envisioned by the Dutch mathematician and cosmographer Andreas Cellarius in 1661. Earth reigns supreme at the centre, circled by a sphere of fire. Mythological figures of the planets race around the Earth on chariots, suspended on clouds. The surrounding ring of the zodiac completes the picture.

40 constellations in addition to Ptolemy's list. Some of these are so far south that the Greeks couldn't see them; others are made from faint stars lying between Ptolemy's main constellations, such as Leo Minor (the small lion) and Lynx – consisting of stars so faint it's said that only the lynx-eyed can spot it!

But Ptolemy's tour-de-force was to sort out the looping path of the planets. The first ingredient in the recipe was that they had to move in circles, at a constant speed, as Aristotle had prescribed. Ptolemy then plagiarized Apollonius's fairground ride theory, with the planets swinging round on small epicycles, as the epicycles themselves – like the spinning cars in a Waltzer – move around the central pivot.

But that didn't quite work. Picking up a clue from Hipparchus, Ptolemy said that the Earth itself wasn't quite at the centre of the celestial Waltzer. That made things better; but not quite good enough for the theory to match the real planets as they paraded through the sky.

Ptolemy's final tweak was to say that the great wheel carrying each planet was moving at a constant rate not around the pivot at its centre; nor around the Earth. To see it moving at its prescribed even speed, you would need to stand at an entirely different point within the circle. Ptolemy called this point the equant.

Ptolemy wasn't just an astronomer. He complemented his astronomy with astrology, in a set of occult books called *Tetrabiblos*. And he brought together everything that was then known about the world in his work *Geographia*.

And later on, Ptolemy had the startling notion of thinking how the geography of the Universe would look, if we could see it from the outside. In his book *Planetary Hypotheses*, he picked up on Eudoxes's old idea that each of the planets (including the Sun and Moon) moved in concentric spheres, nested together like the layers of an onion. The thickness of each layer, according to Ptolemy, depended on the size of the planet's epicycle.

Immediately above the Earth we have the sphere of the Moon, surmounted by Mercury and Venus. The Sun follows. Farther out, we have Mars, Jupiter and Saturn. Finally came the sphere of the fixed stars.

Aristarchus had already pioneered the way to measuring the distance to Moon and the Sun. Now Ptolemy could use his better observations and theory to leapfrog outwards, from one planetary onion skin to the next.

RIGHT *A more realistic, but rather less romantic version of the Ptolemaic System by cosmographer Andreas Cellarius. The small circles (top and bottom) are epicycles – miniature paths that planets took as they circled the Earth on their major orbits. Ptolemy reasoned that epicycles could account for the inexplicable loops that planets made in the sky.*

LEFT *A portrait of Ptolemy by André Thévet, which appeared in his work* Les vrais portraits et vies des hommes illustres *published in 1584. Ptolemy is shown holding a cross-staff for measuring the sky, and pointing to the stars.*

The radius of the Universe, measured outwards from the Earth, was 75 million miles (120 million km) he dramatically proclaimed. That may seem relatively small to us today – inured as we are to astronomers talking of billions of light years – but at the time it was an absolutely mind-boggling concept. Even Archimedes, dreaming of filling the Universe with sand-grains, would have found this a big Cosmos.

It was the first intimation that the Universe is too large for the human brain to grasp intuitively; that we would have to rely on science and maths to fit the Cosmos into our consciousness.

Ptolemy had gone as far as the astronomy of his day could take him. Little was he to know that he had been so thorough in his theory of the Earth-centred Universe that it would stand unchallenged for 1400 years.

And he felt he deserved the reward that's due to astronomers through all the ages: 'Well do I know that I am mortal, a creature of one day. But if my mind follows the winding paths of the stars then my feet no longer rest on Earth, but standing by Zeus himself I take my fill of ambrosia, the divine dish.'

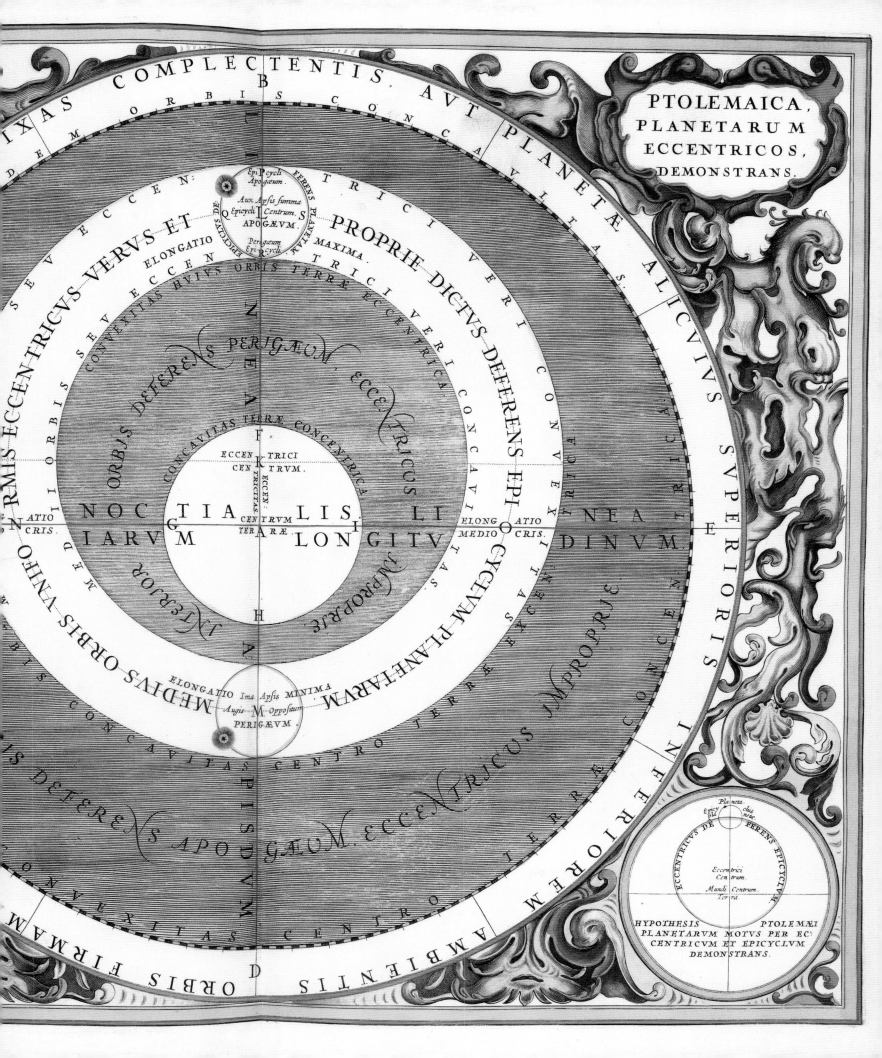

HYPOTHESIS PTOLEMÆI PLANETARUM MOTUS PER EC-CENTRICUM ET EPICYCLUM DEMONSTRANS.

The Earth moves

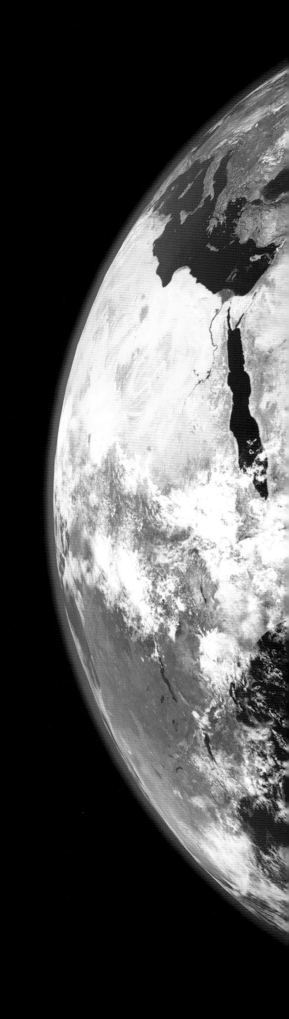

Nine pm, in a small remote town in northern Poland. The soft chimes of a carillon ring out into the gathering dark, peal after peal echoing around the central square. The sound is emanating from a great brick edifice, perched on a rocky outcrop above the restaurant terrace where we've chosen to enjoy a warm late summer evening.

Occasionally, a heavy lorry rumbles through on the byroad from Gdansk to the Russian border. Otherwise, all is still. As the gloom deepens, the first stars emerge above medieval towers on the hill. First, the brilliant trio that makes up the Summer Triangle – the luminous diamond of Vega, softer glowing Altair and amber Deneb. Then the fainter stars making up the outlines of the constellations of the lyre, the eagle and the swan. And finally the glowing band of the Milky Way, arching across the sky.

It's difficult to believe, we muse, that this slumbering giant of a cathedral witnessed the most earth-shattering revolution in the history of astronomy. In a tower that rises to the right of the great church building, an unassuming canon came to astounding conclusion. Though the ground beneath our feet seems solid and immobile, here in Frombork, in 1543, Nicolaus Copernicus made the Earth move.

Copernicus relegated the Earth to be just one of the planets, forcing it to follow a track around the Sun every year. And he put the Earth into a spin, turning round once in 24 hours. Today, we 'know' these as facts, learnt at school and never questioned.

But Copernicus was brought up in a different school, where a stationary Earth was the centre of the Universe. It was a legacy that stretched back to the dawn of time. And a millennium before Copernicus, the geocentric Universe had been given the seal of scientific and religious approval.

Around AD 150, the Greek astronomer Ptolemy had described the workings of the heavens in exact detail. The planets – including the Moon and Sun – move around the Earth in circular paths: to be exact, on the rim of a small circle that's carried round by a larger rotating circle. The beauty of Ptolemy's cosmology lay partly in his circles, which were the perfect geometrical shape. But, more important, it could predict where the planets would be found for centuries into the future.

What were the early Christians to make of this? Their bedrock was the Bible; but the Holy Book doesn't contain an almanac of the planets. Into this confusion came Augustine of Hippo. After a dissolute early life, this great scholar of the fifth century AD became an exemplary saint. St Augustine concluded that God had used some eminent pagans to tell Christians about the physical world: 'it must be said that our authors knew the truth about the nature of the skies.'

After Augustine, the teachings of Ptolemy became fossilised in the Christian tradition for a thousand years. In hindsight, it seems incredible that no-one in Europe questioned them – or even made any serious observations of the sky.

RIGHT *It took a revolution in human thought to appreciate that the Earth is a planet, floating in space.*

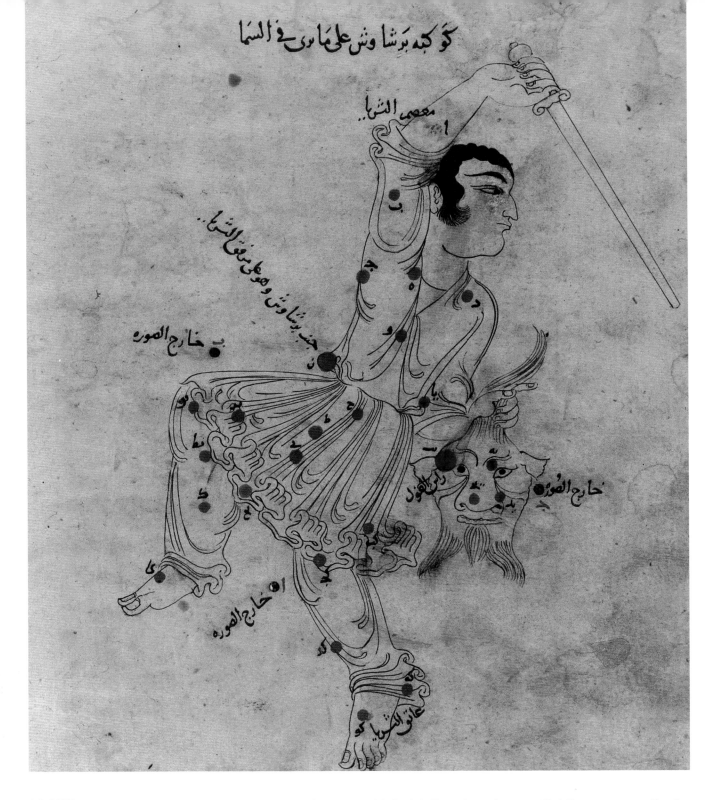

كوكبة برشاوش على مأرئ في السّما

معصم الثريا

خارج الصوره

خارج الصوره

خارج الصوره

ABOVE *Arabic depiction of the constellation of Perseus, the ancient Greek hero who rescued Andromeda. He carries the severed head of the monster Medusa, marked by the star Algol – originally, al-ghul, meaning 'the demon.'*

Allan Chapman, of Oxford, believes it was because Christians were expecting the end of the world at any moment. 'When Christ ascended into heaven, his disciples had understood that he would return to Earth soon after, to bring about the Last Judgment – and then, after a time, to destroy the physical world.'

If the Universe was so ephemeral, and was soon to be annihilated, what was the point of studying it seriously?

Meanwhile, the beating heart of astronomy was to be found far to the East – in the rival civilization of Islam. Unlike the Bible, the Muslim holy book, the Qu'ran, actually exhorts the believer to ponder the heavens. 'In the creation of the heavens and the Earth and in the alternation of the night and the day there are indeed signs for men of understanding.'

Allah had put these signs in the sky for a purpose. The stars helped Muslims to work out the direction to Mecca; while the Sun indicated the five times of day when they must pray. The first appearance of the crescent Moon marked the beginning of a new Islamic month. And – by investigating the heavens – Muslim scholars would literally get closer to knowing the mind of God.

For both practical and religious reasons, the Islamic 'men of understanding' set out to understand how the heavens operate. As their guide, they had Ptolemy's compendious tome: in Greek it was *Mathematike Syntaxis* ('*Mathematical Treatise*'), but the Muslim scholars were so impressed they referred to it as *Almagest* ('*The Greatest*') – the name we still use today.

One thing they didn't question was the Earth's steadfast and unmoving place in the centre of the Universe. Not only did they have Ptolemy's authority; in the Qu'ran, Allah clearly states: 'It is not for the Sun to overtake the Moon; nor can the night outstrip the day. All of them float in an orbit.'

In the centre of Persia (now Iran), astronomer Abd Al-Rahman Al Sufi spent years checking the brightness and colours of the stars. In his *Book of Fixed Stars*, Al Sufi drew up pictures of each of the constellation figures. We can recognise the star-patterns from Greek times, but the characters are dressed in Arabic gear: Andromeda as a Muslim princess and Perseus as a dashing Arab hero, with turban and prominent moustache!

Al Sufi's constellations also included some sky-sights that the great Greek astronomers had missed. Beside his figure of Andromeda is what Al Sufi called 'a little cloud': the first depiction of the Andromeda Galaxy. His *Book of the Fixed Stars* describes another neighbouring galaxy, the Large Magellanic Cloud, as 'the white ox' of the southern Arabs, because it was visible from Yemen but not from Persia.

Intriguingly, he also labels one star in the constellation Perseus al-ghul – 'the demon.' Today we know that Algol 'winks' every three days, as it's obscured by a dimmer star moving in front. Was Al Sufi aware of its strange behaviour? We can only guess…

Like Algol, most of the star names we use today are Arabic. Previously, Ptolemy had generally described stars by their positions in the constellation patterns: but this could get quite cumbersome, such as 'the northernmost of two stars close together over the little shield in the stern.'

By and large, Al Sufi translated the Greek descriptions into Arabic; over the course of the centuries, these came back to the western world in a distorted form. The jewel in the constellation of Piscis Austrinus, for instance, is dubbed by Ptolemy as 'the mouth of the southern fish.' In Arabic, this became fam al-hut al-janub – our star Fomalhaut.

The name of Betelgeuse – the well-known reddish star in Orion – has an odder history. The Arabs saw the surrounding stars as a female figure they called al jauza

BELOW *The leading Persian astronomer Al Sufi studies a star-globe, in this drawing by the sixteenth century German artist Albrecht Dürer. Most of the star names we use today, such as Aldebaran and Vega, are derived from the Arabic descriptions that Al Sufi bestowed on them.*

'the central one', and regarded this star as her hand – yad. But the Arabic letter 'y' looks much the same as their letter 'b'. So this star's name came into Europe as bat al-jauza – hence Betelgeuse – a meaningless phrase that's sometimes mistranslated as 'the armpit of the sacred one'!

Al Sufi took other names from the old traditions of the nomadic Arabs, who gave each star a name that related to their everyday lives, such as the goat, the dog, the cooking fire or the lion. So in the constellation of Cepheus – depicting Andromeda's father – to this day we have a star called Errai, meaning 'the shepherd' in Arabic, and another named Alfirk, 'the flock.'

While Al Sufi stargazed, other Islamic scientists had their heads down, pushing the boundaries of mathematics. The Greeks had been obsessed with geometry; but scholars in the Muslim countries invented trigonometry and algebra. One major breakthrough – quadratic equations – was made by the Persian mathematician Omar Khayyam.

Yes, this is the same Omar Khayyam who's known for his profound poetry. Most famously, he penned the lines: 'The Moving Finger writes; and having Writ,/ Moves on…'

Khayyam was a flamboyant polymath. Born in 1044, he spent his life travelling round to seek out the learned men of his day. In between solving some of the most difficult problems in mathematics, he jotted down his philosophical and mystical thoughts in a set of a thousand four-line poems, the Rubaiyat.

The nineteenth century poet Edward Fitzgerald translated the Rubaiyat of Omar Khayyam into English, and its blend of the exotic and the erotic made it a huge success in Victorian England.

Some of Khayyam's sentiments were distinctly un-Islamic, especially when it came to his beloved tipple, spiced up with female company:

Here with a Loaf of Bread beneath the Bough,

A Flask of Wine, a Book of Verse – and Thou

Beside me singing in the Wilderness -

And Wilderness is Paradise enow.

The multitalented Khayyam was also an astronomer. He measured the length of the year more accurately than anyone before him, and devised a calendar that's theoretically more accurate than the calendar we use in the west today.

Some 300 years later, Ulugh Beg set out on a quest for the ultimate in astronomical precision. He had one great advantage: when he was only 16, his father had given him a whole province to govern. So Ulugh Beg had all the resources he needed to build the greatest observatory the world had ever seen.

The Samarkand observatory must have been a stunning sight for travellers toiling along the Silk Road from Europe to China. A vast drum-shaped building, with a diameter of 160 feet (50 m), it towered 120 feet (37 m) above the top of a rocky knoll. The outside was covered in brightly coloured tiles, depicting the planets' orbits (around the Earth, of course), the stars, deserts, seas and busts of leading scientists from the East and West.

BELOW *Detail from a drawing of Omar Khayyam, the eleventh century Persian scholar, best known for his poetry. But he was also a pioneering mathematician and astronomer: here he is depicted calculating the most accurate calendar of his time.*

Running right through Ulugh Beg's observatory was a huge quadrant: between a pair of high walls, the floor curved upwards from below ground level to the very top of the building, forming a quarter-circle set on edge. Two parallel strips of marble ran along the curving floor. They were inscribed with digits marking degrees around the edge of the circle, like a gargantuan protractor.

The top of the opposite wall was pierced with a small slit, right at the centre of the great stone protractor. This was the sighting device for the astronomer within the building.

It was particularly spectacular when Ulugh Beg used the quadrant to observe the Sun. 'You'd have this dark chamber,' enthuses Allan Chapman, 'where you could open a little opening at the top which would let in a chink of light, so you would have a projection of the solar image on the scale.'

At night, Ulugh Beg or one of his astronomers would sight the stars and planets through the slit, and note their positions from the digits inscribed on the great marble scales. The result was a star catalogue that put Ptolemy's to shame.

Unfortunately, Ulugh Beg's genius in astronomy wasn't matched in politics. While still in his 50s, he was beheaded by his own son. The observatory was destroyed by religious conservatives. It was excavated a century ago, and you can still see the lowest part of the quadrant, the underground trench containing part of the long marble scales.

The nearest we can get to appreciating the scale of the Samarkand observatory is in India. At Jaipur – south of Delhi – an eighteenth century ruler built vast brick instruments for observing the positions of the Sun, Moon, planets and stars. Here you can find the world's biggest astronomical instrument: a sundial with a gnomon that's almost 90 feet (27 m) tall. There are large marble-lined bowls, carefully engraved with scales: an astronomer inside would sight the stars through the intersection of wires running across the top of the bowl.

But by the time the Jaipur observatory was built, it was already well out of date. The western world had caught up with Islamic astronomy and overtaken it. As Allan Chapman remarks, 'These observatories were being built as grand palatial monuments for powerful rulers, but frankly by that date a ship's sextant could have done better.'

The torch of knowledge had begun to pass from Islam to Christianity around the twelfth century, as Christian armies wrested the control of Spain from the Muslims. In Toledo, scholars from France, England and Italy studied with the Muslim scholars in their impressive library. The Christian intelligentsia were astounded. At this time, they had largely lost touch with the sky. For instance, in AD 755 monks in England had recorded that the eclipsed Moon had moved in front of 'a bright star' – a very impressive sight in its own right – without realising that the 'star' was in fact the planet Jupiter.

Now they were hearing about the celestial mechanics of Ptolemy and they discovered the incredible new mathematics that the Islamic scientists had created.

ABOVE *The eighteenth century Jantar Mantar observatory at Jaipur, in India, boasts some of the world's largest astronomical instruments. The sloping wall (right, background) casts the shadow in a vast sundial: its highest point is 90 feet (27 metres) high. The cupola on the top was used to announce eclipses and the beginning of the monsoon. The large brass dials and hemispherical bowls (foreground) were designed to measure the positions and motions of celestial objects.*

Ripples from the new learning spread across Europe, prompting a wave of fresh thinking about the Universe.

The Rector of Paris University was certainly impressed. Jean Buridan was far from being a stereotypical academic: he was involved with any number of women, including the Queen – it was said – and for this last amorous adventure, he narrowly escaped being thrown into the River Seine in a sack!

Nowadays, Buridan is best remembered for his 'logical ass.' If an ass was totally logical, he declared, and it was placed exactly halfway between two equal bales of hay, then it would starve to death because it wouldn't be able to decide which one to eat.

In another 'thought experiment,' Buridan asked if it was possible to decide if the heavens wheel around the Earth once in 24 hours; or if the stars are stationary and it's the Earth that's turning. Everyday experience tells us our world can't be spinning around, or else the atmosphere would be whistling past our ears, and a ball thrown straight upwards would fall some way away from you, because it's left behind as the ground moves. Buridan was one of the first scholars to realise, on the contrary, that both the air and the ball would share in the Earth's rotation and would move with you.

The excitement of this new knowledge and fresh interpretations infected the cultured classes, and the literati of the time were more au fait with astronomy than their counterparts today.

Geoffrey Chaucer, for instance, is best known for his stories told by a motley band of pilgrims, *The Canterbury Tales*. But he also wrote *A Treatise on the Astrolabe*. Here, the great author explains to his young son Lewis, 'lyte Lowis,' how to use the traditional astronomical instrument that he's just given him, configured 'after the latitude of Oxenforde.'

Up to Chaucer's time, Europe was experiencing a trickle of knowledge, filtered through the Arabs. The floodgates opened in 1453. In that year, the great Christian city of Constantinople fell to a Muslim army. For a millennium, Constantinople – now Istanbul – had been the capital of the eastern Roman Empire, the treasure house of all the wisdom of the Romans and the Greeks before them.

Now, scholars fled from Constantinople to Italy, bringing with them books and manuscripts containing the ancient knowledge – including exact copies of Ptolemy's great work, the *Almagest*.

'That bringing in of some of the ancient tradition from the East essentially re-ignited the real renaissance of astronomy,' observes Owen Gingerich from Harvard, 'because if you have no chance of finding and reading and understanding the *Almagest*, you don't have a proper basis for where to go next.'

Johannes Müller, from Königsberg in Germany, was one young man fired by the influx of knowledge. Like many scholars of the time, he's better known by the Latin name of the place he was born: Regiomontanus ('king's mountain'). When his professor died, Regiomontanus was put in charge of the most important astronomical project of the time: translating and shortening the original Greek version of the *Almagest* into a more

BELOW *The Ram Yantra at Jaipur is one of pair of instruments that reveals the Sun's position during the day. The shadow of the top of the central pillar falls onto graduations on the surrounding walls. For continuity, the second Ram Yantra has walls where this one has gaps.*

ABOVE *The Bavarian genius Regiomontanus (born Johannes Müller) was a powerful force in progressing astronomy from medieval mysticism to a mathematical science. He translated Ptolemy's Almagest from the original Greek, and then recalculated all the tables, based on his own observations. Regiomontanus built Germany's first observatory, at Nuremberg, and – according to legend – constructed a wooden eagle and an iron fly which both took to the air. If Regiomontanus hadn't died young (possibly poisoned), he might well have been the first to put the Sun at the centre of the Universe.*

comprehensible Latin text. Regiomontanus also set up his own instruments to observe where the planets lay in the sky – in the process making some useful early measurements on Halley's Comet in 1472. He then went back to the formulae in the *Almagest*, put in his new observations, and recalculated all of Ptolemy's predictions. Whew!

'You could call Regiomontanus the world's first recorded nerd,' elaborates Allan Chapman. 'His passion was calculation. I mean, had he had a computer you could probably never have separated him from it with surgical instruments.'

As a one-man human computer, Regiomontanus was running 'software' based on Ptolemy's view of the Cosmos. The Earth is fixed, and everything else revolves around it. The problem is that three of the planets (Mars, Jupiter and Saturn) sometimes swing backwards in the sky. To accommodate them, Ptolemy put each of them on the rim of a small circle – an epicycle – that's carried round the Earth by a larger circle.

As Regiomontanus cranked out the numbers in his new tables, they predicted the workings of the sky pretty well. They were certainly good enough for the average star-gazer, and for the astrologers making their forecasts.

Did the German genius need to add any bells and whistles to Ptolemy's original scheme, to fit the new observations? Many books on astronomy will tell you that astronomers of the Middle Ages had to add smaller circles onto Ptolemy's epicycles, to match the planets' actual positions... and then they had to add smaller circles again, and again. According to this account, the epicycles became rather like the fleas described by the Victorian mathematician Augustus de Morgan:

Great fleas have little fleas upon their backs to bite 'em,

And little fleas have lesser fleas, and so ad infinitum.

'There's an old legend,' explains Owen Gingerich, 'that gave rise, at some point in the second half of the 19th century, to this mythology that there were epicycles upon epicycles, and that the whole system was about to collapse under its own weight.'

In the late 1960s, Gingerich tried to check out how these extra epicycles would work, by comparing the best astronomical tables of Regiomontanus' time with the paths of the planets according to Ptolemy's simple system. He was expecting to find discrepancies that would reveal the action of the 'lesser fleas.'

But Gingerich was astounded to find the tables matched precisely. He was forced to the conclusion that Regiomontanus and his contemporaries had used only a single epicycle – exactly like Ptolemy. And when he looked further into it, Gingerich found the mathematics of the time was just not up to all the extra complexity: 'So the idea of epicycles on epicycles is entirely mythological.'

Nonetheless, Gingerich is a huge admirer of Regiomontanus's talents: 'If you look at who's doing something in the fifteenth century,' he avers, 'Regiomontanus stands out like a mountain peak. He's not only the leading astronomer in Europe, but also the first astronomical printer.'

Regiomontanus realised that astronomers copying down tables of planetary positions by hand were only human, and would make occasional mistakes. He latched onto the newly invented printing press to mass-produce volumes of error-free tables: he even set up a printing press in his own home.

As Regiomontanus leafed through the original Greek of the *Almagest*, he found that the wondrous book discussed some other ways of laying out the planets' orbits. If he had had the time to follow up this lead, Regiomontanus might have discovered that the Earth actually orbits the Sun, decades before Copernicus.

But in 1575, the Pope called Regiomontanus to Rome, to employ his unique talents in sorting out the calendar. A year later, the German astronomer was dead. It was probably a case of the plague, though there's also a rumour he was poisoned. But his works lived on and may even have saved the life of Christopher Columbus.

On his fourth voyage of exploration across the Atlantic, Columbus's damaged ships ended up on the coast of Jamaica. The mutinous crew ran amok through the hinterland, and the embittered natives eventually refused to provide Columbus with food. Fortunately, Columbus had a copy of Regiomontanus's *Epherimides*. Flicking through the pages, he found that the long-dead German astronomer had predicted a total eclipse of the Moon, in a few days' time on February 29, 1504.

Columbus called the natives together. If they didn't provide him with food, he informed them, his God would make the Moon disappear. And as the Moon rose on February 29, it gradually faded from sight, until nothing was left but a dim blood-red disc.

The terrified natives pleaded with Columbus to make the Moon return, and he agreed on the condition they would provision his crew. As a result of Regiomontanus' 30-year-old predictions, Columbus' men were well fed until a ship arrived to rescue them in June.

Among the students at the University of Cracow, in Poland, who had heard the news of Columbus' incredible first voyage in 1492 was a merchant's son named Mikolaj Kopernik. At college he called himself Nicolaus Copernicus.

Copernicus had been born in 1473 at Torun, on Poland's great river Vistula, the son of a merchant who traded – as his name suggests – in copper. By the age of 10, the young Mikolaj had lost his father. But he was lucky that his unmarried uncle took on the family of four children.

Uncle Lucas Watzenrode was a canon – not an ordained priest, but a business manager for the church's estates. He sent young Mikolaj to university at Cracow, one of Europe's most vibrant cultural cities. Here, Copernicus first encountered the subject that would dominate his life: astronomy. He even bought his own sets of astronomical tables, including one by Regiomontanus, who had died shortly before Copernicus was born.

After Cracow, Copernicus travelled to Italy to study church law at Bologna. The young man was still star-struck, and he managed to get lodgings with the professor of astronomy. We can imagine Copernicus and his prof. sitting up for long evenings discussing the Cosmos as he began his life's work of getting to grips with the motions of the planets.

For such an important figure, though, we have little idea of his personality.

'For me, he was more-or-less a two-dimensional cardboard figure until I went to Uppsala and saw his library,' confides Owen Gingerich, who has studied Copernicus's surviving books around the world. 'Handling the books with his own annotations suddenly gave me a sense of reality – yes, here's a man, here's his library, here are his tracts. But there are so many questions I would like to ask him, like: when you were young did you have a girlfriend, did you enjoy music, did you play an instrument, how did you learn to draw – but we just don't know.'

What we do know is that Copernicus had a critical view of Ptolemy's ancient and revered scheme of the heavens. In particular, the only way that Ptolemy could match the planets' changing speeds across the sky was a mathematical sleight of hand. He said that their speed was constant – not as seen from the Earth, nor even from the centre of their circular orbit (which was not quite the same), but from another point within the circle called the equant. Copernicus was not happy with this fudge, as he wanted perfect motion in the heavens. It was a problem he'd be wrestling for the rest of his life.

After he'd completed his legal studies, Uncle Lucas now had a job for Copernicus in Poland. But the young man was loth to leave the intellectual stimulation of Italy. He persuaded his uncle that he should now qualify in medicine, just up the road in Padua.

In the end, Copernicus spent ten years at three leading universities – but he hadn't actually taken his degree exams. There was a good reason: his student friends would expect him to host a lavish banquet. To save his cash, Copernicus travelled to nearby Ferrara, and took his degree at a university where he knew nobody!

It was time to return to Poland. Lucas Watzenrode was now Bishop of the region of Warmia, a triangular stretch of land in the far north of the country. One of its cor-

ABOVE *Terrified natives of Jamaica cower as Christopher Columbus points out a lunar eclipse that he has predicted – thanks to Regiomontanus's astronomical tables. As a result, the Jamaicans provided Columbus's starving crew with food until they were rescued.*

ners touched the Baltic Sea at the small town of Frombork. Copernicus's post for the rest of his life was to be a canon at Frombork, which he himself called 'this very remote corner of the Earth.'

Today, you come to Frombork by a two-hour drive from Gdansk, over the flat and fertile fields of the Vistula delta. To the left is the nondescript modern town, largely rebuilt after the Soviets flattened it in their advance into Germany at the end of the Second World War. The low cliff to the right is crowned by the red brick walls of the fortified cathedral.

We enter the cathedral compound through a massive gateway, into a grassy courtyard that's dominated by the vast bulk of the church opposite. To our right is an ancient oak tree, its split trunk supported by steel cables: it's affectionately called Copernicus, and was already a century old when the great astronomer moved here.

A squat three-storey tower marks the far left corner of the courtyard. We feel a spine-tingling moment when it sinks in that, in a study on the top floor of this tower, a human being first became convinced that our solid Earth is speeding through space.

But what drove Copernicus to this apparently nonsensical conclusion? Owen Gingerich appeals to the hidden artist within the orthodox canon. 'On the great clock in Strasbourg there's an image with a caption which says 'a true image of Nicolaus Copernicus, based on his own self-portrait.' And if you start to look at Copernicus's manuscripts, it's done with such impeccable draughtsmanship it's really quite extraordinary.

'So this man not only has the talents of an artist, but has the aesthetic sense of an artist. And I think what drove him was the aesthetic sense of unifying things.'

Across the courtyard from Copernicus's tower is the cathedral where he worshipped for so many years. The great brick building has survived miraculously for over six centuries, despite the depradations of the Prussians, the Czechs, the Swedes, the Germans and the Russians. Inside, the cathedral is surprisingly ornate, with high baroque decoration on the altars and every elaborate pillar. The cathedral's most famous son is commemorated only with a simple modern statue.

In 2005, archaeologists found the remains of a body under the floor, which they identified as the great astronomer. With the help of police forensic experts, they reconstructed a face that bears an uncanny – though aged – resemblance to Copernicus's portrait of himself as a young man, with his characteristic long thin nose, narrow cheeks and prominent chin.

That evening, we dine in a pizzeria below the cathedral, charmed by the chimes of its carillon. And we ponder the strange new path that Copernicus had to tread to reach his revolutionary new concept.

'We shouldn't see Copernicus as just literally popping up from nowhere, he's part of a rich tradition,' explains Allan Chapman. 'There was an active and growing late medieval tradition in why the heavens weren't quite behaving themselves, partly through observation, partly through just mathematical delight.'

Copernicus had been reading the book in which the great nerd Regiomontanus had summarised Ptolemy's universe. Each of the outer planets – Mars, Jupiter and

ABOVE The cathedral at Frombork, in northeast Poland, is largely unchanged from the time when Copernicus was a canon there. In the distance, a branch of the river Vistula flows into the Baltic Sea.

Saturn – is carried on a revolving epicycle, which itself moves around on a larger circle. You can envisage the large circle as the rim of a pizza plate, with the Earth as an olive near the centre. A drinking glass gyrating around the edge of the plate is the epicycle, with the planet as a cocktail cherry perched on the rim.

But Regiomontanus had also thought outside the box. He said that, mathematically, this arrangement was the same as making the drinking glass stationary, with the Earth at its centre; and having the centre of the pizza plate waltzing around the rim of the glass. Now the planet is a mark on the edge of the pizza plate.

In a restaurant, such a top-heavy arrangement is bound to lead to disaster. But astronomers of the time believed that the spheres carrying the planets were all weightless, so Regiomontanus' new arrangement was quite plausible.

Copernicus began to play with this idea – and took it one critical step farther. The Polish canon checked out the new arrangement for each of the three outer planets. And he had his first 'eureka moment.'

Imagine we're looking out from central Earth to the edge of the small circle that now encloses our world – the drinking glass in our restaurant model – and we're watching the point on the rim that's the centre for the revolving pizza plate, representing the orbit of Mars. Copernicus discovered that when we look in that direction, we are looking towards the Sun.

He did the same for Jupiter. Again, the Sun lies right in front of us, tracking around the edge of the drinking glass as Jupiter moves around. The same was true of Saturn.

This couldn't just be coincidence. The obvious conclusion was the Sun was actually the point on the rim of the drinking glass that the planets orbited. So, Copernicus reasoned, Mars, Jupiter and Saturn orbited the Sun; later he added Mercury and Venus.

The great Pole was halfway to the idea of a Universe centred on the Sun. But at this time, he still had the Sun – with its attendant planets – circling the Earth.

Even so, the aesthetic side of Copernicus reveled in a new unification that came out of his theory. As far as he knew, Ptolemy didn't put the planets into an orderly system (he actually did, in a later book that Copernicus didn't see). Instead, the ancient Greek calculated the path of one planet at a time, as if the others didn't exist. Copernicus described this system as 'just like someone taking from different places hands, feet, head and other limbs…so that such parts would produce a monster rather than a man.'

The Polish canon's new view of the planets was starting to make more sense. He could relate the size of a planet's orbit to the time it took to circle the Sun.

Astronomers already knew the size of the small circle (the epicycle) compared to the larger circle in a planet's motion. For Jupiter, the big circle was five times larger than the epicycle; for Saturn, 10 times. Now Copernicus had swapped the circles over, and he'd made the smaller circle equal to the distance of the Sun from the Earth. So he could say that Jupiter lies five times as far from the Sun as the Earth, and Saturn

OPPOSITE *This detail of an oil sketch by eighteenth century Polish artist Jan Matejko depicts "The Astronomer Copernicus; or Conversation with God." Atop his tower at Frombork – with the cathedral spires in the background – Copernicus is observing the sky with large wooden rulers (right). The painter has also put Copernicus a century ahead of his time, by placing a telescope at his feet!*

10 times further out. And going in towards the Sun, Mercury lies at only one-third the Earth's distance.

And, from ancient times, astronomers had measured how quickly the planets move around the sky. Mercury is aptly named after the messenger of the gods: it gets around and back to its starting point in just three months; Saturn, the god of old age, lumbers round in 30 years.

For the first time, Copernicus could point out that the further a planet lies from the Sun, the slower it moves.

In his brick tower, looking out over the lagoons of the Vistula's mouth by day and the stars at night, Copernicus wondered if this was all. Did the planets really orbit the Sun, while the Sun moved round the Earth?

And now he had his second eureka moment. By swapping the Sun and the Earth over, he would get a much neater arrangement. If the Earth orbited the Sun as a planet, between Venus and Mars, he would have an arrangement that appealed even more to his artistic sensibility. Our world would travel round the Sun in one year, sitting nicely between the seven months of Venus and the Red Planet's 23-month orbit.

'As Copernicus went through this exercise, he would have suddenly have realised something spectacular,' enthuses Owen Gingerich, 'namely that if he tried to merge it all together in a single system, these ratios are locked together. There's something absolutely beautiful about that arrangement.'

In Copernicus's own words: 'Only in this way, do we find a sure harmonious connection between the size of the orbit and the planet's period of revolution.'

Copernicus wrote up his new vision of the Cosmos in a six-page leaflet, to circulate privately to a few friends. For all its brevity, it was one of the most radical works ever written. Copernicus decreed that the Earth was not the centre of the Universe, but a planet moving round the Sun. He asserted that the looping of the planets in the sky is merely a result of the Earth's own motion. And he declared that the daily wheeling of the stars overhead is caused by our world spinning every 24 hours.

It was a time of revolutions, and not just in astronomy. While Copernicus was rearranging the Universe, Spanish conquistadores were plunging their swords into Central America, the crew of the Portuguese navigator Ferdinand Magellan sailed right round the world, and Martin Luther challenged the moral values of the Church and so launched the Protestant religion.

The world rocked to these great winds of change, but Copernicus's groundbreaking research was still hidden in his study. He was forever polishing his ever-growing manuscript, still trying to make all of the circles in his system turn at a constant rate.

And he was perpetually busy with his professional duties. As a canon, he had to draw up leases for tenants and collect their rents. He also checked on the local coinage, and found some of them were being 'debased' by adding cheaper copper to the silver coins. And when the Teutonic Knights attacked Warmia, Copernicus – temporarily at Olsztin – successfully saved the beseiged town from capture. He also had

domestic problems to sort out, when a new bishop accused Copernicus of keeping his chaste housekeeper as a concubine.

By 1539, Copernicus was approaching 60, and it seemed that his new view of the Cosmos would never be published. Then a miracle occurred – in the shape of a 25-year-old firebrand from Austria.

He called himself Georg Joachim Rheticus, and in his case the Latin name was not just for show. His father had been executed for sorcery, and young Georg was not allowed to use the family surname. A loud, forceful and energetic personality, he was later in life to be sentenced to 101 years exile for a gay relationship with a young stu-

ABOVE *Copernicus's Tower still stands in the cathedral close at Frombork. On the top floor of this building, Nicolaus Copernicus started the revolution that demoted the Earth from the centre of the Universe to a mere planet.*

dent. In 1539, Rheticus had heard rumours of Copernicus's new ideas, and made a pilgrimage from Germany to Frombork.

Rheticus was immediately won over by Copernicus's Sun-centred Universe. And the ageing canon must have been flattered by the younger man's enthusiasm. Even so, it took Rheticus another two years to persuade Copernicus to commit his great work to print. The volume would comprise six 'books.'

The first book is the most important to modern eyes: it contains the arguments for a Sun-centred Universe. To convince the skeptics, Copernicus plays up the Sun's importance: 'In the midst of all dwells the Sun. For what better place could you find for the lamp in this exquisite temple, where it can illuminate everything at the same time?'

The other five books include mathematical theories, a star catalogue and detailed calculations on the motions of the Sun, Moon and planets – using Copernicus's complicated method of making them move in circles at a constant speed.

No printer in Warmia could handle the complex tables and diagrams, so Rheticus arranged for the book to be printed in Nuremberg. The printer kept Copernicus posted with proofs, which the Polish canon corrected and sent back.

But in December 1542, Copernicus suffered a stroke. His health gradually deteriorated. The last printed pages arrived in front of Copernicus's eyes on May 24, 1543; by the end of the day he was dead.

He was – we hope – too ill to fully understand the anonymous introduction that arrived with the completed pages that day. While Rheticus was away from Nuremberg, the proofreader at the printers – a clergyman named Andreas Osiander – inserted a preface stating that since an astronomer 'cannot in any way attain to the true causes, he will adopt whatever suppositions enable the motions to be computed correctly …these hypotheses need not be true nor even probable.'

The introduction negated Copernicus's conviction that the Earth actually does circle the Sun, by stating it was simply a convenient mathematical device. Osiander's words certainly left astronomers of the time confused. According to Gingerich, some scholars believed that Copernicus's most important contribution was simply to make all the planetary circles turn at a constant pace.

And those who did understand didn't see any conflict with their religious beliefs. 'Nobody was burned,' comments Allan Chapman. 'The idea of the Earth moving around the Sun wasn't that offensive, it just seemed to defy common sense.'

Over the next century, though, the arguments would heat up to incandescence. The fuse had been lit by the great astronomer who had amassed powerful arguments that the Sun stands still and the Earth is in motion.

Copernicus's book was entitled *De Revolutionibus Orbium Coelestium Libri Sex* (*Six Books on the Revolutions of the Heavenly Spheres*), usually shortened to *De Revolutionibus*. And it was indeed a work that would lead to the greatest revolution in the history of astronomy. This smoldering powder keg would eventually explode the ancient scheme of the Cosmos.

OPPOSITE *Copernicus's great book* De Revolutionibus *contains a diagram that overturns all previous conceptions of the Universe. The central position is occupied not by the Earth, but by the Sun (Sol). It's orbited by Mercury and Venus, and – further out – Mars, Jupiter, Saturn and the sphere of the stars. Most audaciously, Copernicus places the Earth, orbited by the Moon, between Venus and Mars: our world is demoted to a mere planet.*

net, in quo terram cum orbe lunari tanquam epicyclo contineri
diximus. Quinto loco Venus nono menfe reducitur. Sextum
deniq; locum Mercurius tenet, octuaginta dierum fpacio circù
currens. In medio uero omnium refidet Sol. Quis enim in hoc

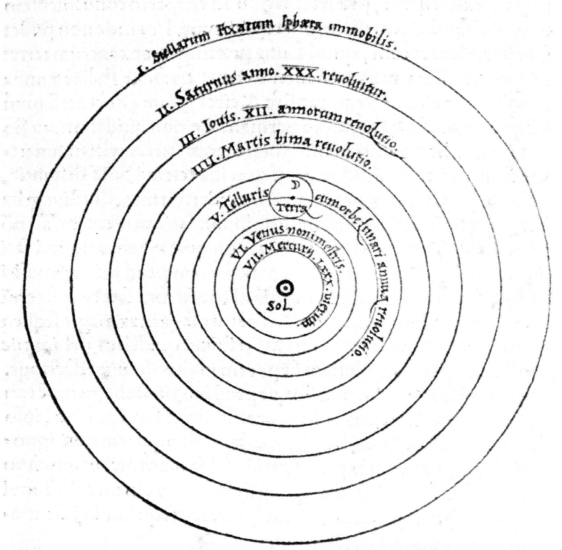

pulcherimo templo lampadem hanc in alio uel meliori loco po
neret, quàm unde totum fimul pofsit illuminare? Siquidem non
inepte quidam lucernam mundi, alij mentem, alij rectorem uo-
cant. Trimegiftus uifibilem Deum, Sophoclis Electra intuentè
omnia. Ita profecto tanquam in folio re gali Sol refidens circum
agentem gubernat Aftrorum familiam. Tellus quoq; minime
fraudatur lunari minifterio, fed ut Ariftoteles de animalibus
ait, maximã Luna cũ terra cognatiõe habet. Concipit interea à
Sole terra, & impregnatur annuo partu. Inuenimus igitur fub
hac

The New Solar System

The alchemist leaned back from the esoteric books on his bench, stretched and looked around. To one side, an assistant was pumping at the bellows, raising the flames that heated a crucible containing a few drops of precious molten metal. To the other, another servant was carefully watching a graceful curved glass alembic, dripping a thin stream of pale yellow fluid.

But it was now time for dinner. And this alchemist was no impoverished seeker after truth, not a beggar for his crust as he was for knowledge.

The 25-year-old Tycho Brahe was a clansman of the most powerful families of the day. His relatives controlled the Kingdom of Denmark – and, to all intents and purposes, controlled the King himself.

Tycho's laboratory was an underground annex to his uncle's grand house, the monastery of Herrevad. It was no longer a place of worship. All over northern Europe in the sixteenth century, monarchs and powerful courtiers had wrested monasteries from the Church and turned then into palatial homes.

Tycho imperiously cast his glance around. Everything seemed to be in order. Climbing the stairs to ground level, he had to pull his coat tightly around himself as he was struck by the chill evening air.

For several days, the weather had been muggy, unseasonably warm and cloudy. Tonight, though, the sky was clear and brilliant. And as Tycho glanced up, he was in for an even bigger surprise.

And 'surprise' is a definite understatement. What Tycho Brahe saw on that evening, November 11, 1572, would transform his life. More importantly, it would change forever our view of the heavens, and humankind's understanding of the Universe.

Almost directly overhead, Tycho saw a dazzling point of light. It lay in the constellation of Cassiopeia, a W-shape of five stars supposed to represent an ancient Queen of Ethiopia. But the new object far outshone these ancient stars.

'Amazed and as if astonished and stupefied,' he later recalled, 'I was led into such perplexity by the unbelievability of the thing that I began to doubt the faith of my own eyes.'

He turned to the servants behind him and told them to direct their gaze heavenwards. They agreed there was a bright star overhead. But the overbearing Tycho was perhaps too used to his servants being 'yes-men.' So he stopped a passing carriage and asked the country people on board. 'These people shouted out that they saw that huge star, which had never been noticed.'

The new star struck at the very roots of science. Tycho lived in a time when scholars believed the ancient Greeks Aristotle and Ptolemy, who had taught that the Earth was the centre of the Universe, circled by the Moon, Sun, planets and stars. More recently, Nicolaus Copernicus had proposed that the Sun should take centre stage, in place of the Earth.

But whatever astronomers thought of Copernicus, they still followed Aristotle in one of his key teachings. There was a fundamental difference between what happens

BELOW *Though remembered today as a great astronomer, Tycho Brahe devoted much of his time to alchemy, in an underground laboratory like this depiction by the sixteenth century Flemish artist Johannes Stradanus.*

Within the portrait:

RVDER · VLETANDER
LONGER · RØNNOR
ROSENKRANS · TROLLER
AXELLSØNNER · LONGER
MARCKEMAN · ROSENSPAR
KABBELER · STORMVASE
GVLDENSTEREN · AXELLSØNNER

NON HABERI · SED ESSE

EFFIGIES TYCHONIS BRAHE OTTONIDIS DANI
DÑI DE KNVDSTRVP ET ARCIS VRANIENBVRG IN
INSVLA HELLISPONTI DANICI HVENNA FVNDATORIS
INSTRVMENTORVMQ' ASTRONOMICORVM IN EADEM
DISPOSITARVM INVENTORIS ET STRVCTORIS
ÆTATIS SVÆ ANNO 40. ANNO DÑI. 1586. COMPI..

LEFT *Portrait of Tycho Brahe at the age of 40, published as the frontispiece of his Astronomical Letters. The bridge of his nose appears slightly distended, because it is metal replacement for flesh severed in a duel. Fortunately Tycho's eyes survived the encounter, as he became the greatest observational astronomer of his era.*

on Earth and events in the sky. In our own surroundings things change, and things decay. But the heavens are perfect: the stars and planets move in their pre-ordained paths, for eternity. Nothing in the Universe can ever change.

Yet here was a change in the sky that was absolutely blatant. Only one explanation would fit with Aristotle's dogma: what appeared to be a star was actually in the Earth's atmosphere – with its inconstant parade of clouds, rainbows and other meteorological phenomena.

Tycho was the man to check this out. Although he was largely engrossed in alchemy at this time, Tycho was at heart an astronomer: he even called alchemy 'terrestrial astronomy,' believing that the reactions that fizzed or fumed in his flasks were a mirror of the interactions between the planets and the stars.

In his excitement over the new star, Tycho immediately brought out his cross-staff, a simple wooden device that he could use to measure the distance between two stars. In turn, he checked the celestial interloper's distance from each of the five major stars of Cassiopeia.

Over the next few nights, he repeated the observations over and over again, at all times of night. During the dark hours, Cassiopeia and its brilliant star wheeled around the Pole Star, from high in the east to due south, and ending up low in the northwest as the dawn came up.

And however hard Tycho tried to measure any movement at all, the star remained obstinately fixed in its place in the constellation. As Tycho persistently came up with a negative result, he realised just what it meant...

From the time of the Greeks, astronomers had been aware that perspective shifts the position of the Moon in the sky, during the course of the night. (This is in addition to the Moon's own motion around the sky, once a month.) Our changing viewpoint on the Moon makes it appear to move, just as the position of a fly on your window seems to shift against the background if you view it with one eye, and then the other. If the new star lay in the Earth's atmosphere, it should seem to shift overnight even more than the Moon.

Tycho knew what parallax the Moon shows: it appears to shift by almost twice its own width. In contrast, the new star moved by less than one-thirtieth the width of the Moon. So it must lie dozens of times farther away than the Moon – out with the planets or the stars. Because the star didn't move like the nearer planets, Tycho argued it must lie out in the realms of the stars: in Aristotle's cosmology, the uppermost or eighth sphere. In Tycho's own words, 'this new star is not located in the upper regions of the air just under the lunar orb, nor in any place closer to Earth, but far above the sphere of the Moon in the eighth sphere...'

So the science of Tycho's time was wrong. The stars were not constant and immortal. Like things on Earth, the celestial realm was subject to change. As Tycho's star dimmed over the next year, it was clear that even the heavens could experience decay.

Tycho had proved that the heavens and the Earth were not different. His measurement of the new star of 1572 was the thin end of a wedge of increasingly accurate measurement that would plumb deeper and deeper into space, and eventually prove that our world is but a small and insignificant part of a vast Universe, where laws of nature rule planets, stars, galaxies and the Cosmos alike.

The great Dane had begun a lifetime of scientific research, which would lead astronomers to break the mould of the heavens. And Tycho led a correspondingly unconventional and larger than life existence, right from his birth...

The Danish astronomer was born in 1546, just three years after Copernicus had died – promoting on his death-bed the mind-blowing concept that the Earth circles the Sun. His father, Otte Brahe, was a leading noble in the Danish court and his mother, Beate Bille, came from an equally aristocratic family.

Beate gave birth to twins. The first-born was christened Tyge: when he became a serious scholar around the age of 15, Tyge followed the usual custom and turned his name into a suitably academic Latin form, as Tycho. The second twin died. But Tycho never forgot his twin brother, and later said that he achieved so much in life because he was doing the work of two.

When Tycho was a year and half old, his life took an even stranger turn. While his parents were away from their home at Knutstorp Castle (now in southern Sweden, but then part of the Kingdom of Denmark), his uncle Jørgen Brahe came to call – and kidnapped the young Tycho. It meant that Tycho enjoyed all the attention of an only child, for his uncle and aunt never had any children of their own.

This curious twist of family history helped to create the scholar who probed the depths of the Universe. His foster mother, Inger Oxe, came from an intellectual family.

OPPOSITE *Tycho steps out of his underground laboratory, and spots a brilliant new star in the sky. This later engraving may dramatise the impact on the bystanders, but the supernova of 1572 was pivotal in overthrowing ancient concepts of the Universe.*

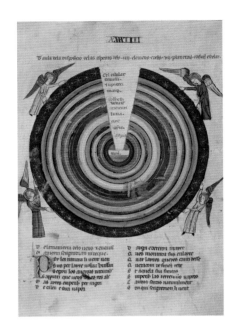

ABOVE *Until Tycho's time, scholars believed the planets were carried on solid crystalline spheres, centred on the Earth. Religious authorities taught that angels provided the propulsion. Tycho's observations would prove that solid spheres couldn't exist in the heavens.*

While Tycho's natural parents would have propelled him into a career as a diplomat and courtier, Inger pushed the young Tycho towards a university education.

In those days, students didn't specialise. Tycho had to study everything from mathematics to Hebrew, and music to rhetoric. But the study of the heavens was the subject that caught his soul: both astronomy – and astrology.

At the time, most scholars believed that the heavens must have an influence on the Earth. Certainly the Sun affects our lives, the passing seasons and the crops in the field. The Moon was responsible for the tides, though no-one knew the reason. Tycho himself argued that God would hardly have created the vast and complex wonders of the heavens unless they had some role to play in the life of his greatest creation, human beings.

Allan Chapman puts Tycho's belief in astrology into perspective. 'Ancient wisdom attributed particular virtues to planets: for instance, Jupiter was jolly, Mars was rapacious and violent. Particular houses of the Zodiac had particular characteristics. So he wasn't being superstitious or silly. Astrology was part of their culture in the same way that motor cars are part of ours.'

Tycho's passion was kindled by an eclipse in 1560; and fired by a close approach of the planets Jupiter and Saturn three years later. And he then did something unknown for a young aristocrat. He got hold of a large pair of wooden compasses, and used them to make his own measurements of the two celestial bodies as they closed with each other.

To his surprise, he found that all the predictions of this event were wrong. Astronomers who relied on the wisdom of the ancients – the old Ptolemaic theory where the Sun circles the Earth – were out by a whole month. Predictions based on Copernicus's new theory of a Sun-based Universe were wrong by several days.

The 16-year-old Tycho, armed with only a rough instrument, had shown up the predictions of his olders and betters. It was bad for both astronomy and astrology. 'What's needed,' he concluded, 'is a long-term project with the aim of mapping the heavens, conducted from a single location over a period of several years.'

Before that dream could come true, his private life was in for some tumultuous times. Late in 1566, Tycho got into a quarrel with a fellow student. According to Owen Gingerich, 'they were arguing about who was the better mathematician. It was a Christmas party, and I think they must both have been terribly drunk. They went outside to have a duel….'

The fight ended when his opponent's broadsword slashed the bridge off Tycho's nose. 'It must have been a really bloody mess,' opines Gingerich. Allan Chapman concurs: 'A few years ago I was talking to a facial repair surgeon, a Mr Ninian Peckitt, who said there'd have been a massive haemorrhage because two major arteries come through the skull there…'

Tycho fixed his nose with a metal bridge. According to tradition, it was made of silver and gold. But Gingerich points out 'When the Czechs exhumed him in 1901, there was a green stain on the skull. So there's some copper involved in this.'

Perhaps his golden nose had a copper clip to hold it in place. Or perhaps Tycho sported his gold nose for special occasions, while wearing a lightweight copper nose for everyday. Unfortunately, Tycho lived too early to benefit from Mr Peckitt's latest prosthetic, a titanium bridge under the skin – which he calls the 'Tycho Brahe nose.'

By the end of 1571, Tycho was living at Herrevad monastery with an uncle (not his father's brother, who had brought him up, but his mother's brother). And now he experienced an upheaval of a pleasanter kind. Tycho fell in love.

But this was an ill-starred romance, at least from his relatives' point of view. Kirsten Jørgensdatter was a commoner – the daughter of the local clergyman. Technically, the aristocratic Tycho could not marry her. But there was a loophole in Danish law. If a woman lived openly in a man's house for three winters, eating and drinking and sleeping together, and wearing the keys to the house at her belt, then she could be called his wife.

The catch was that she could not appear in public with her husband, wasn't entitled to use her husband's surname and couldn't inherit any of his property. Their

BELOW *Uraniborg was Tycho's observatory and manor house rolled into one. The domes and conical roofs housed his astronomical instruments, while his alchemical laboratory was sited in the basement. The central block was more mundane, containing the living quarters for Tycho's family, staff and guests.*

IOANNIS KEPPLERI,
Mathematici Cæſarei
hanc Imaginem,
ARGENTORATENSI BIBLIOTHECÆ
Conſecr.

children would be in the same position. Tycho's son would not bear the noble family name: instead of being Tyge Brahe, like his father, he would be known merely as Tyge Tygesen ('Tyge, son of Tyge').

But Tycho was young, obstinate and infatuated. Kirsten became his common-law wife. And they were to enjoy a long and happy relationship, through to Tycho's death.

As the aristocratic Danishman was wooing his pastor's daughter, a baby boy was born to a very different family, 515 miles (830 km) to the south. Unknown of course to Tycho at the time, this event would lead to the most significant partnership in the history of astronomy.

Johannes Kepler was born into a thoroughly dysfunctional family in the southern German city of Weil der Stadt. By Kepler's own reckoning, his father was 'vicious, inflexible, quarrelsome and doomed to a bad end' – he eventually abandoned his family to become a mercenary soldier. Kepler was no kinder in his description of his mother: 'thin, swarthy, gossiping and of a bad disposition.' She dabbled in herbal medicines, and would later be tried as a witch.

To be fair, Kepler was as uncompromisingly harsh on himself: 'That man has in every way a dog-like nature… He is malicious and bites people with his sarcasms. He hates many people exceedingly and they avoid him, but his masters are fond of him. He has a dog-like horror of baths, tinctures and lotions.'

Perhaps Kepler should have taken his hygiene and medications more seriously. He was constantly in poor health. His skin was always breaking out in boils, he suffered constant problems with his digestion, experienced foul headaches and endured piles throughout his life. Oddly, for an astronomer, he also had bad eyesight.

Kepler was less than a year old when the new star erupted in Cassiopeia, and astounded Tycho Brahe. Today, we know it was not 'new' at all, but the end of the road for an old star. Some 7,500 light years away from us, a tiny dense star was ripping gas from a companion rather similar to our Sun. Unfortunately, the small star faced a severe weight problem. Once a white dwarf like this reaches a certain weight – 40 per cent heavier than the Sun – it becomes unstable. The white dwarf explodes as a cosmic nuclear weapon: imagine a hydrogen bomb that's more massive than the Sun!

A small-scale exploding star is called a nova – from the Latin for 'new'. A star explosion on the scale of Tycho's merits the name supernova.

Today, all that's left of Tycho's exploded star is a fireball of ultra-hot gases. This supernova remnant has now grown so tenuous that it's hardly visible, even to the world's biggest telescopes. Yet it has a special place in one of our lives (Nigel's), because I studied Tycho's supernova remnant from Cambridge using some of the world's best radio telescopes. At one time I could claim to be the world expert on this distant cloud of gas! My research showed that the fireball is a tangle of magnetism, which still shines brilliantly if you observe the radio waves it's emitting.

In Tycho's time, no-one knew that the supernova represented death on a galactic scale. But astrologically, it was a deadly omen. It transformed the 'W' of Cassiopeia into a cross: the sign of Doomsday. Tycho foresaw 'war, rebellion, murder, disease, death

LEFT *Johannes Kepler was the greatest mathematical astronomer of his day. Totally convinced that the Sun lies at the centre of the Universe, he was doggedly determined to investigate how the planets – including the Earth – move around this central body. Kepler would eventually break with the ancient conviction that heavenly bodies travel in circles, and determine that the planets follow egg-shaped elliptical paths.*

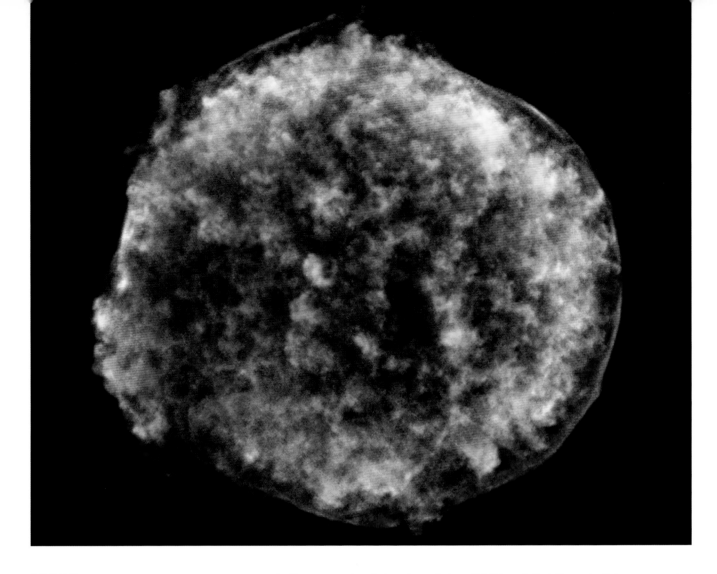

ABOVE *The fireball from Tycho's exploding star of 1572 has faded from view, as seen with ordinary telescopes. But it still shines brilliantly, in this view from NASA's Chandra satellite observatory, which detects X-rays from space. The tangled cloud of superhot gas – at a temperature of 18 million°F (10 million°C) – now stretches over 15 light years of space.*

and all manner of unfortunate and dreadful things.' And the celestial omen worried King Frederick II of Denmark. He called Tycho to court. The two men got on well, and Frederick sent Tycho all over Europe as a royal agent, to bring in the best art and artifacts to adorn the new castle, Kronborg, he was building on the edge of the strait that separates Denmark from Sweden, at Elsinore.

Yes, this was indeed the castle that Shakespeare was to immortalize in *Hamlet*, only 25 years later. As the Bard researched his play, he must surely have come to hear of the great astronomer. And there are some tantalizing hints to Tycho in the play.

Hamlet starts with three soldiers on a platform before the castle of Elsinore. Bernardo recalls that a ghostly figure had appeared the previous night: 'When yon same star that's westward from the pole/ Had made his course to illume that part of heaven/ Where now it burns…'

Is this burning star Tycho's brilliant supernova? As a boy, Shakespeare certainly could have seen the new star of 1572, and its dramatic debut in the heavens would provide a simile for the arrival of the glowing Ghost of Hamlet's father on the battlements.

The British astronomer Cecilia Payne-Gaposhkin was one of the first to make another connection: according to Shakespeare, Hamlet was a graduate of Wittenberg – Tycho's old university.

Even more intriguing are the names Rosencrantz and Guildenstern as courtiers in the play. Shakespeare probably met two Danish noblemen, Frederick Rosenkrantz and Knud Gyldenstierne when they were envoys to London: these gentlemen were both relatives of Tycho, and they must surely have spoken to the ever-curious playwright about Denmark's most flamboyant character.

American astronomer Peter Usher has more recently suggested that the whole play is an allegory of the contest between astronomers of the time. The wicked king Claudius represents the Earth-centred view of the Universe, as proposed by the ancient Greek astronomer Ptolemy: his full name was Claudius Ptolemaeus. Rosencrantz and Guildenstern stand in for their relative Tycho. And Hamlet acts the role of the new Sun-centred astronomy of Copernicus.

Again, there's good evidence that Shakespeare personally knew the main champion of Copernicus in England, Thomas Digges. He in fact went further than Copernicus. Digges said that the stars are not fixed to a sphere that surrounds the Sun

LEFT *Richard Burton stars as Hamlet – holding Yorick's skull – in a 1953 production of Shakespeare's play at the Old Vic in London. Hamlet was the Prince of Denmark, and there are many clues in the play that the Bard knew of Tycho Brahe and his astronomical ideas.*

LEFT *Stjerneborg was Tycho's second observatory, largely built underground for stability and to keep the astronomers warm! Above ground, two observers (lower right) are measuring the separation of stars with Tycho's large wooden sextant.*

– like the shell of a walnut – but are scattered at greater and greater distances, out to infinity. Maybe that's what Hamlet means when he says 'I could be bounded in a nutshell and count myself a king of infinite space…'

Whatever thoughts the events in Denmark had excited in Shakespeare's mind, by 1575 King Frederick was certainly ensconced in Kronberg Castle at Elsinore. And he had something on his mind – which wasn't a ghost. He'd heard that his esteemed envoy and friend Tycho Brahe was planning to emigrate to Germany.

The King had intended to bring Tycho more into court circles, and serve the Danish crown like his ancestors. He'd even offered him a range of rich estates. But Tycho had tactfully refused. Now Frederick was hearing that Tycho was 'displeased with society here, customary forms and the whole rubbish.' Instead, the maverick astronomer wanted just one thing, a place that would suit 'a student interested in learned studies.'

Frederick's gaze fell on a small island just off the coast opposite. And the King saw the possibility of granting Tycho just what he most desired….

The island of Hven (now part of Sweden, and usually spelt Ven), rises from the sound between Denmark and the Swedish coast like a flat cake. Steep cliffs, about 100 feet (30 m) high, are all you see as you approach on the ferry from Copenhagen. From the small port, you climb to the plateau, and the island is almost flat – a gentle amble of a couple of hours will encompass all its sights. These days, you have to avoid the tourists speeding past on their bright-yellow hire bikes, but the island is still as quiet and agricultural as it was in Tycho's time. Now, you'll find goats producing cheese for the tourist trade, and durum wheat for pasta; back then, it was simpler husbandry, with the farmers growing oats and rye, and tending cattle for their milk and beef.

Tycho landed on Hven in 1576, and claimed it for his own. The farmers – instead of having to undertake military service for the King – were entirely at Tycho's beck and call. As Lord of Hven, the noble-born Tycho Brahe created his own little kingdom – devoted not to matters of this world, but dedicated to studying the Universe.

At the highest point of island, Tycho designed a building the like of which the world had never seen. It was a manor house, a laboratory, an observatory and a research institute rolled into one. And its appearance wouldn't be out of place in Disneyland. Two ornate porches opened to the east and west, while circular wings to the north and south were crowned with wooden platforms and conical domes. Curving Dutch gables on the central block were topped with more domes, while a weathervane of Pegasus – the winged horse – reared proudly on the summit.

The interior echoed the arrangement of the Universe. In the basement was Tycho's alchemical laboratory. On the first floor, the domes on the wooden platforms each contained an instrument for observing the heavens. Between, on the ground floor, Tycho located his living quarters.

Right in the centre, Tycho erected an elaborate fountain, which he turned on to impress important visitors 'with a water-carrying figure rotating around and throwing water in all directions.' But the focus of the ground floor was his circular study – which Tycho called the 'museum,' as it was the room dedicated to the muse of astronomy, Urania. Here he housed his massive library of 3000 books, along with portraits of ancient astronomers and globes depicting the Earth and the sky.

Tycho's most treasured possession was a set of wooden measuring rulers that had belonged to Copernicus himself. Tycho had sent an expedition to Frombork, to measure exactly the latitude where Copernicus had made his observations. The canons were so impressed that they donated this invaluable relic of their great predecessor to his natural successor.

Abutting the mansion were aviaries; and it was surrounded by geometrically laid-out gardens of herbs and fruit trees. These were enclosed in a great earthern rampart, with entrances at each of the points of the compass. Tycho christened this magnificent edifice Uraniborg – the Castle of Urania.

The celestial castle buzzed with all kinds of people. They included Tycho's wife and their growing number of children – except when the nobility visited, and Kirsten had to be hidden away. There were the usual servants, as in any manor house. But Tycho's staff also included his astronomical observers and the craftsmen who made his painstakingly accurate instruments for observing the sky. And there were the scientists – his own staff of students, and visiting academics who came from all over Europe for the honour of working with the world's most famous scientist.

By today's standards, the daily routine started early. Apart from the observers – who were working all night – everyone was up and about by four o'clock in the morning. Breakfast could be porridge or herring, washed down with warm beer. The main meal of the day came in the evening, before everyone went to bed around at 8 pm.

And Tycho's supper menu matched his extravagant personality. Over the course of the evening, the servants would bring in two or three 'sets' of food, each consisting of several courses. Soup, venison, goose liver, carp, lamb, rye bread, pike, crayfish, chicken pate, sugar cakes, almond sweets – all these and more might appear in a single evening. Most people drank several quarts of beer a day; Tycho himself would have wine with his supper.

Everyone was expected to provide some entertainment, by singing, playing the lute or declaiming poetry. For a time, Tycho kept a dwarf, called Jeppe, to amuse the company. Tycho certainly enjoyed a varied entourage: a few years earlier, he had kept a pet elk, but it drank too much beer, fell down the steps and died.

For all the superficial trappings of a nobleman's dining hall, the table talk at Uraniborg was more akin to the discussions at an academic college. Tycho's noble brothers on the

OPPOSITE *This illustration of Tycho's quadrant seems to show an oversized version of the great astronomer compared to the other figures! In fact, the details within the curve were painted on a wall in Uraniborg where the scale for this precision instrument was engraved. An astronomer (far right) is observing a star, while a second (lower right) reads the exact time, and a third (lower left) records the observation. In the painting, we see Tycho and the three floors of Uraniborg, from the astronomical instruments on the roof to the alchemical furnaces in the basement. In contrast to all this activity, Tycho's hound remains somnolent!*

ABOVE *A nineteenth century woodcut depicts a rather more 'scientific' medieval view of the Cosmos, where the angels have been replaced by clockwork. Though Tycho wouldn't have bought into the details of this engraving, he concurred with the idea that the Earth was central and fixed.*

mainland would be discussing the wars now raging between Protestants and Catholics or the English Queen Elizabeth's defeat of the Spanish Armada. But the Lord of Hven would be dominating a conversation that ranged from the efficacy of his latest alchemical brew to the astronomical ideas and controversies coursing through Europe.

The most important debate concerned the Earth. Did it reside in the midst of the Universe, as Ptolemy had asserted? Or was Copernicus correct, and the Earth was but one of the planets that circled the Sun?

Tycho – for all his ground-breaking ideas in science – was old-fashioned on this issue. He couldn't believe that the massive Earth was rushing through space. He couldn't even believe it was spinning on its axis, once a day. Like many other scholars, Tycho was influenced by the Bible. In the Old Testament, Joshua had commanded the Sun to stand still – not the Earth.

But Tycho had scientific evidence up his sleeve as well. If the Earth goes around the Sun, then our viewpoint on the stars must change during the year, and they should appear to shift slightly backwards and forwards. And Tycho's precision measurements showed no signs of this parallax: the stars stood resolutely fixed.

Tycho could draw one of two conclusions. Either the Earth was stationary in the centre of the Cosmos. Or the stars lay so far away that the parallax shift was too small for him to detect. The latter would turn out to be the correct answer: the stars do shift slightly, in fact, because of parallax – but it would another 250 years before anyone would have the equipment sensitive enough to measure the tiny movement in the sky.

Tycho refused to believe the stars could be so far off. If Copernicus was right, then the distance to the great celestial sphere carrying the stars must be a thousand times further away than the most distant planet, Saturn. Tycho could not believe that God would create a Universe with so much wasted space.

And he apparently had more science to back him up. The brighter stars in the sky appear to be bigger than the smaller ones. No-one takes that seriously today, because we know it's an optical illusion: starlight spreads out in the retina – the light-sensitive screen in the back of our eyes – and the light from a bright star spreads further than the light of a dim star.

But Tycho thought the effect was real. In that case, a brilliant star like Sirius would be thousands of times wider than the Earth. And that was just ridiculous…

On the other hand, Tycho wouldn't just go along with Ptolemy. His predictions for the planets were highly complicated and not very reliable; and in fact Tycho used Copernicus's calculations, even though he didn't believe the Polish canon's ideas!

For years, Tycho wrestled with the problem. He was concerned to avoid 'both the mathematical absurdity of Ptolemy and the physical absurdity of Copernicus.' First, Tycho conceded that Venus and Mercury must circle the Sun. That's the easiest way to explain why they never lie far from the Sun in the sky. And eventually, he worked out his own complete theory for the Solar System.

In the Tychonic system, the Earth lies steadfastly in the centre of the Universe: it doesn't even spin. The Moon circles around the Earth. Farther out, the Sun is also circling the Earth. And the key ingredient is that all the other planets are circling the Sun, as it travels around our world.

Mercury and Venus are the Sun's close companions. The other planets – Mars, Jupiter and Saturn – follow large circular paths around the Sun that can take them right round to the other side of the Earth. That's when we see these planets high in the sky at night.

Tycho was delighted. He had once and for all solved the problem of the planets. There was just one thing that kept troubling him…

BELOW *Tycho was proud of his own vision of the Universe, which steered a course between Ptolemy's Earth-centred cosmos and the Sun-centred cosmology of Copernicus. The Sun circles the central Earth which remains stationary; meanwhile, the other planets orbit the Sun.*

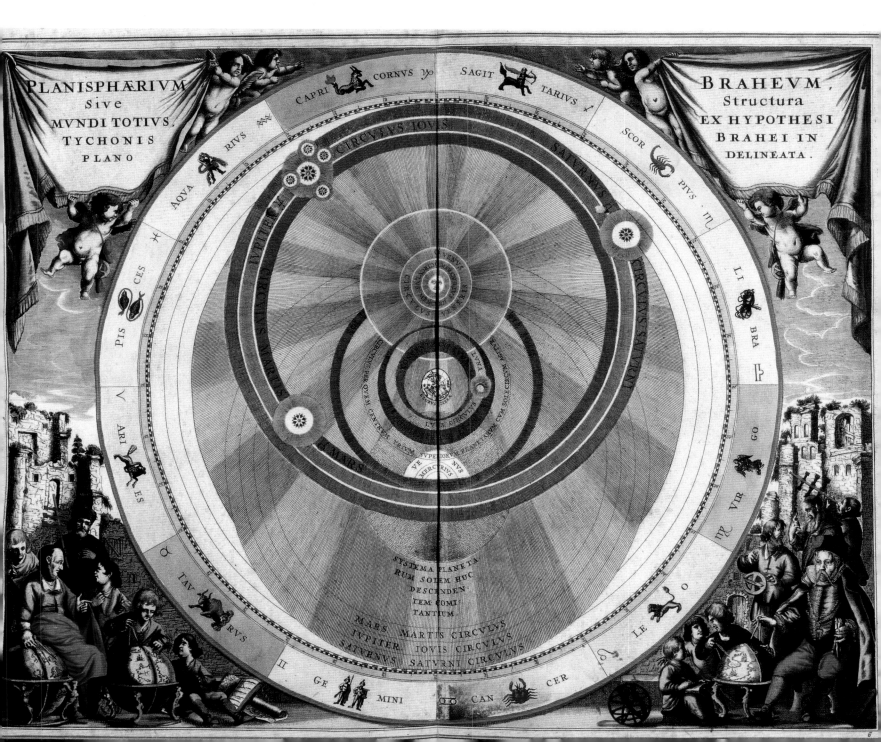

ABOVE *A contemporary broadsheet shows the brilliant comet of 1577 moving through the sky, from Sagittarius (right) past Delphinus (the dolphin) to Pegasus (the flying horse, at left). As the comet recedes, its tail shrinks. This comet would have smashed through the crystalline spheres – proving they didn't really exist.*

Everyone knew that the planets traveled around the heavens in circles because they were carried on huge revolving spheres, made of transparent crystal. Some paintings of the time even depict angels pushing each of the spheres around, to keep the planets moving at their God-appointed speed.

For Ptolemy, the spheres were all centred on the Earth. For Copernicus, the Sun was at the centre. In either case, the spheres nestled inside each other, like a set of well-rounded Russian dolls. But in Tycho's new system, the sphere holding Mars actually cut through the sphere which carried the Sun around the Earth.

Tycho again wrestled with the problem. At first, he 'could not bring myself to allow this ridiculous penetration of the orbs, so that for some time this, my own discovery, was suspect to me.' And the answer came from an unexpected direction, as Tycho started to write a book on an event that happened several years earlier, just after his arrival on Hven.

On November 13, 1577, the new Lord of the Manor had been fishing in one of the estate's fishponds when he saw another 'new star.' As it had grown dark, he'd seen it had a tail: it wasn't a star, but a comet. Tycho had measured the comet's position, proving that it flew above the Moon's orbit. This was more evidence that the heavens were not perfect, but suffered changes just like the Earth.

As he wrote up his book on the comet in 1586, Tycho worked out the path it must have followed through the Cosmos. To his surprise, he found it had traveled right across the paths of Mercury and Venus. The comet should have smashed into the crystalline spheres of the two planets – but there was no sign of a cosmic collision. Clearly, the spheres were not actually solid after all.

Tycho had removed the last obstacle to his theory. He rushed his new vision of the Cosmos into print as the final chapter of his book on the comet of 1577 –

De mundi aetherei recentioribus phaenomenis ('*On the most recent phenomena of the ethereal world*').

His excitement, it would turn out, was misplaced. Tycho's model of the Cosmos would not stand the test of time. But the comet of 1577 would have two lasting influences on the history of astronomy. First, Tycho's destruction of the crystalline spheres meant that astronomers would have to find some other way to keep the planets moving in regular paths.

And, far to the south of Hven, a mother was showing her young son this same comet. Katharine Kepler – for all the faults that her son was to list – was interested in natural phenomena. And she made a special point of waking six-year-old Johannes and taking him out to view the celestial sight. Three years later, she made sure that he saw an eclipse of the Moon. These sky sights ignited a lifelong passion for astronomy.

Meanwhile, Tycho was not resting on his laurels. Even 'working as hard as two men,' he had plenty more to achieve. First, he wanted to chart the positions of the stars in the sky with the utmost accuracy. And he needed to make the final measurements of the planets that would prove his new vision of the Cosmos.

The master of the heavens kept his instrument-makers busy, constructing ever more accurate sextants, quadrants and armillaries. The wooden platforms on Uraniborg were too small, too wobbly and too exposed for the new precision instruments.

So Tycho constructed a second observatory, just outside the banks enclosing Uraniborg. It was called Stjerneborg – 'castle of the stars'. This was partly underground, so the instruments would be more stable and the observers protected from the weather. Tycho also included a small fire and beds for extra warmth.

With his great instruments, Tycho – or increasingly, his assistants – could carry on the work on the great star atlas. His goal was to chart the positions of the 1000 brightest stars to an accuracy of one arcminute. An arcminute is one-sixtieth of a degree in the sky. We can put that in context by saying it's about one-thirtieth the width of the Full Moon. But that wouldn't be fair to Tycho, because the brilliant Moon always looks bigger in the sky than it really is. Instead, think of the three bright stars that make up the Belt of Orion, the constellation representing the legendary hunter. Tycho was aiming to measure positions accurate to one-hundredth the distance between each of the Belt stars – and without a telescope!

And Tycho introduced another new method into science: to measure everything several times over, with different instruments, and take the average of all the results. As he became satisfied with each star's position, he not only entered it into his growing catalogue of stars: he positioned it on his great star globe.

This globe was his proudest possession, and the 5 foot (1.5 m) sphere dominated his 'museum.' He had commissioned it from Germany back in 1570. But in the years it took to arrive at Uraniborg, the wooden sphere had warped. Tycho was not content with anything less than perfection. He filed down the bulges, filled the gaps, glued on layers of parchment – and then covered it all with brass sheets, engraved with lines

BELOW *Tycho's Large Armillary was the most accurate astronomical instrument constructed before the invention of the telescope. With it, he could measure the position of stars or planets with a precision of one-hundredth the diameter of the Moon.*

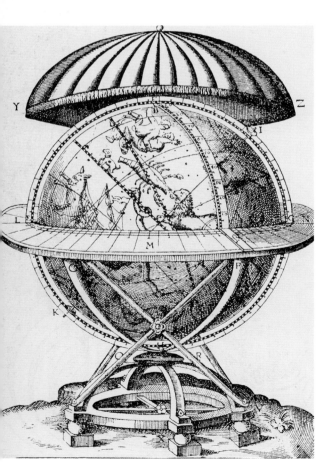

representing celestial latitude and longitude. By the time he left Hven, it would be studded with 777 precisely located stars.

Finally, Tycho was checking out the movements of the planets – and, in particular, how Mars was behaving. According to the Tychonic theory of the Cosmos, Mars should come closer to Earth than any other planet. If he could measure the parallax of Mars, he could prove his theory was correct. With hindsight, we now know that Mars's parallax is less than Tycho could measure – but his focus on the Red Planet would bear unexpected fruit after his death.

Tycho's observations and writing were interrupted all too often by visitors, drawn to meet this legendary figure. The most welcome was his sister, Sophie. She was unusually fascinated by knowledge for a woman of her time. When she was 17, she'd helped her big brother Tycho observe an eclipse of the Moon. Later, she explored alchemy as well with a view to devising new medicines. Her concern echoed Tycho's. His interest in alchemy was fuelled not by a desire to turn lead to gold, but to cure illnesses.

The most surprising visitor, perhaps, was King James VI of Scotland – later King James I of England. In 1589, he became engaged to the 15-year-old Princess Anne of Denmark. A delegation of noblemen, headed by Tycho's brother Steen Brahe, set off with Anne for Scotland, but bad weather forced them to take shelter in Norway. James was so impatient to meet up with his young bride that he took to the storm-tossed seas, and married her in Oslo.

The next spring, James and Anne visited Hven. Tycho gave them a guided tour of his strange learned domain, and threw a banquet that was lavish even by his standards. The King was so impressed that he presented Tycho with a pair of English mastiff dogs; and James later personally penned a poem (in Latin) that ended:

What Phaethon dared was by Apollo done,

Who ruled the fiery horses of the Sun.

More Tycho doth, he rules the stars above,

And is Urania's favourite, and love.

And there's an intriguing Shakespeare connection here, too. He wrote his last play, The Tempest, for King James. And the plot involves a shipload of noblemen (including a king), who are caught in a fierce storm and wrecked on an island that's ruled by a sorcerer, Prospero, who has acquired magical powers through his great library and his learning. Was the Bard inspired by the tale of the King's visit to Hven?

Meanwhile, Tycho's relationship with his own royal family was not as assured. The sympathetic King Frederick had just died. Fortunately for the great astronomer, the heir to the throne was still a minor, and the kingdom was ruled by four senior noblemen – all related to Tycho. They approved his annual income, and even increased it.

At that time, Tycho was receiving – and spending – a phenomenal amount of money. As well as Hven, he'd been given estates throughout Denmark and Norway to provide funding for his research activities. Tycho's income amounted to one per cent of the total crown revenues – the same percentage that NASA receives today from the United States federal budget!

Are to be sold at the whit horse in pope head Alley by Iohn Sudbury and George Humble

For Johannes Kepler, in Germany, money was a much more pressing issue. His teachers had recognized his brilliance early on, and he was now studying at the University of Tübingen. Kepler was still fascinated by astronomy – part of his university course – but he intended to make a career in the Protestant church. His superiors, though, had other plans, which would mean a salary that he could scarcely live on.

Still, Kepler had little choice. Now, as throughout his life, his actions were driven by the fierce conflict that was raging throughout central Europe between the established Roman Catholic Church and the new Protestant religions.

The Catholics had set up a high profile school in the Protestant city of Graz, in what is now Austria. To ensure the children of Graz wouldn't be exposed to Catholic doctrine, the Protestants had established a rival school. They wanted the best teachers in each subject: and the best mathematician to hand was undoubtedly Johannes Kepler.

During his lonely stint in Graz, Kepler had plenty of time on his hands. He didn't make friends easily; and he was now free of his theological studies. His mind turned to astronomy. Unlike Tycho, Kepler didn't believe the Earth was the centre of the Universe. He was a committed follower of Nicolaus Copernicus.

Copernicus's great book *De Revolutionibus* had been published 50 years earlier, and had created a huge stir. As a mathematician, Kepler realized the book was not just a wild new stab at re-ordering the cosmos, but was based on thorough mathematics.

Copernicus had worked out the distances of the planets from the central Sun. But why those particular distances? (We now know that these distances were all too small, but the proportions of the orbits were correct.)

Kepler believed that God had not chosen the planets' distances at random. So he set out to find the pattern. After months of failure, the answer came in flash, one day in July 1595, as he lectured to his students. On the blackboard, Kepler drew an equilateral triangle, with three equal sides, and a circle that ran through its three corners. Out of the blue, he wondered what would happen if he drew another circle, inside the triangle and touching its sides. When he measured up the two circles, their sizes were in the same proportion as the orbits of the planets Jupiter and Saturn.

Kepler was elated. But he couldn't make the geometry work for the other planets, using triangles or other flat shapes like a square or a hexagon. Then he had his second insight: he should try three dimensional shapes to fit between the planets. Since the time of the ancient Greeks, mathematicians had known there are just five 'perfect' solid shapes, where every face is the same as every other face. The most familiar is the cube, with six square faces. There's also the pyramid-shaped tetrahedron (with four triangular faces), and shapes with eight, 12 and 20 faces.

When Kepler nested these shapes together, with a sphere separating each from the next one out, he ended up with six spheres. And their sizes matched the orbits of the six planets. Kepler broke into tears. 'I believe it was by divine ordinance that I obtained by chance that which previously I could not reach by any pains.'

We know today that there's no truth behind Kepler's model. But even now, planetary scientists disagree as to why the orbits are just the size they are.

In 1596, Kepler was busy with two projects. One was wooing a wealthy miller's daughter, Barbara Mueller. The other was publishing a book on his breakthrough, under the title *Mysterium Cosmographicum* ('*Mystery of the Cosmos*').

From the beginning, Kepler had intended to a send a copy to Tycho on Hven. But the book never reached the Danish astronomer in his island kingdom. Tycho was now faced with the greatest crisis of his career. The heir to the throne, Christian IV, had just been crowned. And a new broom was about to sweep through Denmark.

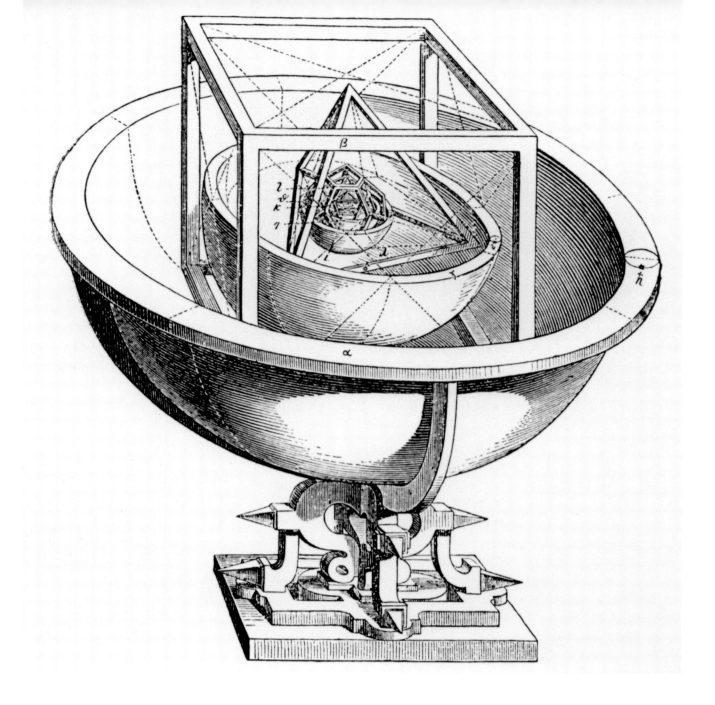

ABOVE *In his book* Mysterium Cosmographicum *('Mystery of the Cosmos'), Kepler placed the planets on nested spheres, separated by the five 'perfect' solids. The outermost sphere here, representing Saturn, is separated by the outline of a cube from Jupiter's smaller sphere; which is separated from Mars by a tetrahedron, and so on. Kepler intended to fabricate a silver version of this design, to use as a punchbowl for dispensing drinks!*

Christian was determined to centralise power around the throne, and that meant he had to break the power of the old nobles. Tycho, with his lack of a power-base and his unconventional marriage, was an obvious target. Christian's antagonism to Tycho was political, not intellectual. Christian was a very cultured monarch, who invited the English composer John Dowland to Elsinore. He later built an observatory on top of a round tower in Copenhagen, with an interior spiral ramp that the King could drive up in his horse-and-carriage: it survives today as the oldest working observatory in the world.

Within a month of Christian's coronation, Tycho lost his biggest revenue-earner, his estates in Norway. Next, he was held on suspicion of various trumped-up charges, including complaints that he had oppressed the villagers of Hven and let the church fall into ruin. Another prosecutor stated that Tycho 'for 18 years has not taken Holy Eucharist but has lived an evil life with a mistress.'

Time was running out for Tycho and his family. Because his children couldn't inherit his family estates, he'd sold his possessions for cash that he could pass to his sons and daughters. The new moral judgment meant there was a danger the state would prevent them inheriting anything at all.

Almost overnight, in the spring of 1597, Tycho packed up what instruments he could – including the great star-globe – and sailed from Hven for ever.

Today, there's virtually nothing left of what had been the world's greatest observatory. One of us (Nigel) visited Hven in 1971, and was dispirited to find only 'a depression 100 feet (30 m) long in which are revealed a well and the foundations of a room,' as I wrote at the time. As soon as Tycho departed, his fellow islanders had used Uraniborg as a convenient quarry for bricks and stones to rebuild their own dwellings in a finer style.

In recent years, a small but bustling tourist centre has grown up. The outline of Uraniborg is marked by a neatly clipped box hedge, one of the ramparts has been rebuilt, and the garden has been replanted with herbs, flowers and fruits that are known to have grown in this region in Tycho's time. The ruins of Stjerneborg now contain an audio-visual display of just how Tycho and his observers would have worked through a long, cold Danish night.

Just outside the rampart mounds, a nineteenth century church contains a museum to the great Danish astronomer. Despite all that we knew about Tycho, we felt a thrill run up our spines when we were confronted with exact replicas of two of his bigger instruments: a wooden sextant as long as your outstretched arms; and a steel quadrant, taller than either of us, with its scales engraved to a fraction of a millimeter.

Tycho had to leave this great steel instrument behind – but he travelled with as much as he could carry. He first went to Germany, which he knew from his student days, and where he had many friends. Here, Tycho received a copy of Kepler's new book. The great Dane was not impressed with the theory – it didn't accord with his own Universe, centred on the Earth. But he recognised the work of a mathematical genius. And he started asking around about the young German who had written *Mysterium Cosmographicum.*

Tycho's sights were now set on Prague. The Holy Roman Emperor, Rudolph II, was desperate to have a dependable astrologer. The post of Imperial Mathematician was vacant: and Tycho quickly secured the position.

But the atmosphere of Prague couldn't have been further removed from the rural idyll of Hven. Packed inside its wall in a bend of the Vltava River, the city swarmed with artisans, merchants and students. The great Charles Bridge, fortified with a magnificent gateway, led to mighty Prague Castle crowning the hill opposite. More a palace than a fortification, it's reputed to be the largest castle in Europe. Here, Rudolph surrounded himself with artists, astrologers, alchemists and magicians.

The new Imperial Mathematician quickly persuaded the star-struck emperor to put him up in an estate outside the capital, called Benatky. This was the venue for the fateful meeting of the two most important living astronomers, in February 1600.

For Kepler, too, was on the move, bedeviled by the religious schisms of his country. A new ruler had assumed power in Inner Austria, and had imposed the

BELOW *The Rundetaarn (Round Tower) in Copenhagen is the world's oldest functioning observatory: it was built by Tycho's nemesis, King Christian IV, in 1637. The top platform originally housed copies of Tycho's instruments; now there's a telescope for public viewing of the stars and planets.*

BELOW *This ornate 'Temple of Astronomy' forms the frontispiece of the Rudolphine Tables, the great work that encapsulates Tycho's observations and Kepler's theoretical predictions of the planetary motions. The astronomers in the temple include Hipparchus, Copernicus and Tycho. Below, the central panel depicts Tycho's island of Hven; with, to the left, Kepler sitting at a table.*

Catholic faith even on Protestant strongholds like Graz. Kepler was banned from any professional post, whether teaching or preaching.

This would give him time to fathom the remaining harmonies that lay tantalisingly hidden in the heavens. Having worked out the arrangement of the planets, Kepler now wanted to know how – and why – they move in the way that they do. For that, he needed access to the best measurements he could lay his hands on. And that meant seeking out Tycho Brahe in Prague.

Though united by a common dedication to the Universe, the two characters could hardly have been more different. Tycho was large, loud and domineering, his distinctive long moustaches contributing to his aristocratic mien. Kepler was thin, withdrawn and devoid of small-talk – though forceful to the point of rudeness when it came to promoting his ideas. Both men needed each other, but each was suspicious, too.

Tycho, at the age of 54, knew that he didn't have enough time left to analyse all of the observations he had accrued, on top of his continuing programme of observing the stars and planets. He needed Kepler's mathematical skills, and his comparative youth. On the other hand, he knew that Kepler was a confirmed disciple of Copernicus, and might use Tycho's observations to try to disprove Tycho's own model of the Universe.

Indeed, Kepler desperately wanted Tycho's observations, to investigate the motion of the planets, in a Sun-centred universe. But he also was also worried that Tycho would treat him more as a hired-hand than a scientist, in preparing a vast new table of planetary predictions that Tycho had promised the Emperor, with the flattering title The Rudolphine Tables.

Nonetheless, the deal was struck. Soon afterwards, Rudolph moved his Imperial Mathematician back to the city, where he dwelt in various great houses on the hill beside Prague Castle, as befitted his noble status. Kepler, by contrast, lived in a succession of small premises in the crowded streets within the city walls – one of which still survives today, along the busy tourist route from Charles Bridge to Old Town Square.

The final large instruments from Hven eventually arrived, and Tycho set them up in the Belvedere, an ornate summerhouse on a terrace behind the Castle. The Belvedere stands today, looking down on the river Vltava and the busy centre of Prague, though the instruments have long since disappeared.

But Tycho was not to last long. In October 1601, he attended a banquet given by Count Rozmberk. He felt a call of nature – but couldn't leave the table until his host rose. According to Tycho's own account, his bladder burst under the strain. Certainly, he suffered 11 days of agony and delirium, before his life slipped away, muttering 'May I not have lived in vain.' This story is so riveting that it's been repeated down the centuries. But is it true?

'Of course not,' expostulates Owen Gingerich, professor of the history of astronomy at Harvard. 'That would have killed him straight away.'

'Instead, he suffered a slow painful death,' Gingerich continues, 'probably of some urinary infection that prevented him from urinating, and he couldn't keep himself from continuing to drink.'

Allan Chapman, of the University of Oxford agrees: 'It seems to have been some kind of urinary stoppage.' One possibility is an enlarged prostate gland, pressing on Tycho's urinary tubes.

Is it possible to find out, after all this time? We certainly know where his body lies, in the fantastically spired Tyn Church, the cathedral right on Old Town Square in Prague. To this day, you can see the deeply incised gravestone – now set upright against a pillar – with its carving of Tycho, his hand set on a globe.

In 1901, the worthies of Prague opened the grave, to find the remains of a man with a damaged nose. They clipped off part of his beard. Ninety years later, with the end of Communist rule in Czechoslovakia, a small sample of the beard was gifted to the Danish government. Swedish scientists have made a minute forensic examination of the hairs, and found an abnormal amount of mercury, in the section that grew a couple of days before he died.

Owen Gingerich explains: 'he may well have self-medicated himself, because he was known for making alchemical potions and those containing mercury were often prescribed for urinary ailments.'

Within two days, Kepler was appointed Imperial Mathematician. In the time he had free between calculating the massive Rudolphine Tables, writing a book on optics and drawing up astrological predictions for the Emperor, Kepler was now hot on the trail of the question that had perplexed astronomers since the time of the Greeks: the strange motions of the planets as they parade around the sky.

Others had been there before, of course. Ptolemy had described planetary circles arranged around the Earth, while Copernicus favoured circles centred on the Sun. But Johannes Kepler came in with a totally new approach in the history of science.

'Kepler was very unusual as what I would call the world's first astrophysicist,' explains Owen Gingerich. Kepler wasn't content just to draw pretty geometrical diagrams showing how the planets move; he wanted to get to the underlying reason as to why they move.

In the old days – when the Earth was the centre of things – it was quite reasonable to think a patient angel was pushing the outermost sphere that carries the stars, once in 24 hours. This motion is then transmitted downwards, slipping a bit all the way, through the spheres of the planets – including the Sun – right down to the Moon.

But in the new Universe, the outer heavens only seemed to whirl around, because the Earth itself was spinning round once a day. Now the mechanics of the Universe looked different. The Sun was stationary at the centre. The innermost planet, Mercury, moved fastest; and the outermost world, Saturn, was the loiterer of the planetary system. Gingerich says that the conclusion, to Kepler, was inescapable: 'Because Mercury is the one that's going the fastest, then the driving force has to be coming from the centre – from the Sun.'

As Kepler was settling into his job in Prague, he came across an intriguing book written by an English doctor, William Gilbert. Later to be physician to Queen Elizabeth, the unmarried Gilbert had thrown all his spare time into investigating magnetism.

ABOVE *The ornate Tyn church dominates Prague's Old Town Square. The medieval exterior has stood unchanged since the time of Tycho and Kepler: note that the two towers are not symmetrical, supposedly to symbolize masculine and feminine principles.*

BELOW *Tycho's tombstone in Prague's Tyn church, carved in red marble, shows the great astronomer dressed in a nobleman's full armour, with his hand on a globe. Above his grave, a plaque declares 'Non faces nec opes, sola artis sceptra perennant,' meaning – in Latin – 'Neither power nor wealth, only Art and Science will endure.'*

He was the first scientist to suggest that the Earth itself is a huge magnet, with the power to influence compass needles. He'd even spent a fortune on building a giant spherical magnet that he could use to check out his theories.

Gilbert believed, like Copernicus and Kepler, that the planets orbit the Sun. He suggested that magnetic forces might be involved in their motions…

The idea fell neatly into place in Kepler's mind. He was already convinced that the Sun had to be the centre of the planetary system because it was so much bigger than anything else – using Tycho's results, it could contain over 100 Earths – and it had the power to generate light and heat that we could feel. The pallid Moon and the other feeble planets, visible only at night, paled into insignificance. Now, magnetism could provide the power that the Sun used to drive the planets around.

This first hint of physics in the Universe appalled Kepler's colleagues across Europe. His old professor in Germany, Michael Maestlin, wrote in indignation: 'I think that one should leave physical causes out of the account… The calculation demands astronomical bases in the field of geometry and arithmetic.'

But Kepler, the self-styled dog, had the bone firmly between his teeth – and he wasn't about to let go. He focused in on Mars, because the Red Planet had always been the most difficult world to fit into a neat planetary orbit.

Kepler had told Tycho, before his death, that he would sort out Mars in a week. In fact, his 'warfare with Mars' took many years. He spelt out the 900 pages of calculation in his book *Astronomia Nova* ('*New astronomy*'), and breaks off in the middle to address the reader with the words: 'If this wearisome method has filled you with loathing, it should more properly fill you with compassion for me, as I have gone through it at least seventy times…'

Fortunately, Prague was providing some relief from the tedium. Things weren't going so well at home – Kepler was now calling Barbara 'simple of mind and fat of body' – but he was now going out and about more. Emperor Rudolph's fascination with art, architecture, music, science and mysticism had made the city the cultural centre of northern Europe. And Kepler was broadening his social horizons, making new friends and developing a good sense of humour – which he often used on himself. Above all, people were carried away by his enthusiasm, whether for the motion of the planets, the date that Jesus Christ was born, an exploding star that appeared in the constellation Ophiuchus in 1604, or the shape of a snowflake.

But the work on Mars was going badly. He just couldn't massage the path of Mars around the Sun to fit Tycho's exacting observations.

Then Kepler had an insight that altered everything. Copernicus had said that the Earth's circular orbit around the Sun isn't properly centred, so that the Earth is slightly closer to the Sun in January. But everyone had assumed that the Earth travels round this circle at a constant speed.

If the Earth is closer to the Sun in January, however, then it must soak up more of the Sun's mysterious power, the astrophysicist in Kepler deduced. And more power should impel the Earth to go faster. So Kepler sat down to work out just how the Earth's

speed changes as it lapped the Sun every year. Without the aid of a modern computer, Kepler created geometrical drawings to represent the baffling motions in the sky. Now, he drew a circle to represent the orbit of the Earth, with the Sun slightly off-centre. For the Earth's position each day, he drew a line from our planet to the Sun – like thin slices through a cosmic pizza.

Some of these pizza slices were long and thin, when the Earth was far from the Sun and moving slowly. Six months later, the slices were shorter but broader, as the Earth hurried through its near-point to the Sun. That's when Kepler stumbled across an amazing fact. Despite their different shapes, each of the daily pizza slices had the same area. The effect of the Earth's changing distance and its varying speed exactly cancel out.

'He says, 'behold, it's a miracle!" enthuses Gingerich. Johannes Kepler had broken free of the chains that had kept astronomers bound since the time of the

ABOVE *English scientist William Gilbert demonstrates the power of static electricity to Queen Elizabeth I. Gilbert's pioneering experiments on magnetism convinced Kepler that the Sun must be exerting a similar force, that drives the planets around their orbits.*

ABOVE *The planet Mars – seen here by the Hubble Space Telescope in a view that would have astonished Tycho or Kepler – held the key to unraveling the orbits of the planets. Of the easily seen planets, Mars follows the least circular track around the Sun. From Tycho's meticulous observations, Kepler proved that Mars's orbit is an egg-shaped ellipse.*

Greeks – their devotion to the idea that the planets move in perfect circles and at a constant speed. At this time, Kepler still had the Earth moving in a circle, but he had proved that the mysterious power of the Sun was driving it at a speed that constantly varied, depending on its distance.

Now Kepler went back to Mars with renewed enthusiasm. He put Mars in a circular orbit, obeying the same speeding law as the Earth. Things fitted much better. But there was still a problem. When he calculated the Red Planet's motion round the sky, west to east, he got a slightly different result from working out its slight wobble north and south.

The discrepancy was tiny – just eight arcminutes, or one-quarter the diameter of the Moon. Before Tycho Brahe, no-one had measured Mars accurately enough to have even spotted it. Kepler was in a fix. Did he take the easy way out and gloss over the observations? No; though he was not a keen observer himself, Kepler trusted Tycho's observations more than he trusted the theory.

It was back to the drawing board – literally. After another couple of years, Kepler realized he could only make it all hang together if he squashed the circular orbit of

Mars slightly – as he later described it, 'as though I were squeezing a German sausage in the middle.'

Squashed circles weren't the easiest of shapes to work with, though, so Kepler tried his hand at using an ellipse to make the maths easier. To his astonishment, the ellipse gave a perfect fit. 'It was in incredible leap of imagination,' Gingerich explains, 'which happens to be right.'

An ellipse is the shape you see when you look at a circle at an angle. The ancient Greek mathematician Apollonius had first investigated ellipses around 200 BC. Ironically, Apollonius had also been responsible for setting up the epicyclic 'circles-on-circles' machinery to explain how the planets move – the model that Kepler now threw in the garbage can.

Kepler had the key to the heavens, at last, and could open the secret vault of the planets' motions. The treasure trove that had been signposted by Copernicus and Tycho was now revealed. Every planet travels around the Sun on an elliptical path, at a variable speed that's governed by the 'law of areas.'

Today, Kepler's Laws (these two, plus another published later in his life) are the basis not just of astronomy, but of space travel through the Solar System. And they apply just as much to the newly discovered planets that orbit other stars and may be the abodes of life.

But in his lifetime Kepler didn't receive the praise he deserved. He published his results in the immense *Astronomia Nova*, so full of his calculations that his main conclusions were hidden under the mass of detail. It would take the genius of Isaac Newton to realize what the hard-working dog of Prague had achieved.

Meanwhile, astronomy was about to take a totally unexpected turn. Just after *Astronomia Nova* was published in 1609, a friend knocked on Kepler's door in great excitement. An Italian astronomer named Galileo Galilei, he reported, had used an 'optick tube' to observe the sky. And apparently Galileo had discovered four new planets.

Kepler was astounded. Four new planets would destroy his vision of the Solar System (in fact, they were the large moons of Jupiter). Yet he immediately saw the potential of an instrument with a power far beyond even Tycho Brahe's eyes.

With the intellectual power of Copernicus and the meticulous observations of Tycho behind him, Kepler had overthrown the old theory of the planets. That was as far as the naked eye could take astronomy. Kepler was visionary enough to realise that a whole new era of astronomical discovery was about to open up.

BELOW *Glowing red in the Sun's setting rays, a statue of Johannes Kepler at the Griffith Observatory, Los Angeles, gazes out into the cosmos. Kepler oversaw the demise of the old Earth-centred Universe; he was also one of the midwives of the new cosmic perspective opened up by the telescope.*

Beyond the Human Eye

On a sticky hot day in Florence, it's a relief to enter the air-conditioned Museum for the History of Science down by the River Arno. Our mission is to track down the very beginning of modern astronomy: two telescopes made by the great Galileo.

Up the stairs, past brass astrolabes, into the first of the Galileo rooms. And amidst the gleaming metal, polished wood and shining glass, there's something distinctly organic. It resembles a bent, dried-up stick.

Look more closely – and it's a finger. Distinctly, a mummified human finger, pointing upwards from a glass dish that holds it reverentially, like the relic of a saint.

According to the caption, this is the finger of Galileo himself. To be precise, it's the middle finger of his right hand.

What is Galileo's finger telling us? It's tempting to think that he's making a rather rude gesture towards the Church Fathers who censored his words, stifled his thoughts and imprisoned him. Or perhaps he's beckoning us on, to the great new discoveries about the Universe revealed by his telescopes?

Despite what people often say, Galileo didn't actually invent the telescope. He wasn't even the first person to turn it towards the sky. What was so important is that Galileo was more than just a celestial sightseer, of the 'been there, got the T-shirt variety. He realized that these new sky sights would change the way we understand the Universe – and all in one miraculous year, from the autumn of 1609 to autumn 1610.

Without the telescope, we would still be stuck today with the astronomy of the sixteenth century, with good evidence the Earth circles the Sun but no proof; with no idea of what the stars are; with no concept of galaxies, black holes and the Big Bang. This great revolution was started not just by Galileo's discoveries, but by the sheer force of his personality.

RIGHT *Galileo delighted in showing his visitors the wonders of the cosmos through his telescopes. They included the English poet John Milton, whose famous poem,* Paradise Lost, *contains references to the great man's telescopes.*

OPPOSITE *Two of Galileo's early 'optick tubes' peer upwards to the mysteries of the sky. The lenses were tiny: only about 1.5 inches (4cm) across. It is a tribute to the astronomer that he was able to observe anything at all, and – moreover – to record his findings scientifically.*

ABOVE *From his workshop, Dutch spectacle-maker Hans Lipperhey uses two lenses to bring ships in the port closer. His invention would have important implications for the military, as well as for astronomy.*

'Galileo has a hard face in the portraits of him as a young man,' says Allan Chapman of Oxford University. 'It's a very tough, very powerful face. Frankly, you wouldn't want to meet him on a dark night.'

We've met his skeletal finger in a sunny Florence day, though, and it leads us to the next room, and the two ancient telescopes made by Galileo. They are rather quaint wooden tubes, with a small lens at each end, one stylishly covered in leather.

Telescopes like these overturned the Universe. Yet there's a long-running dispute about who invented them. About 250 years earlier, craftsmen in Venice had begun making small discs of curved glass that people could support in a frame in front of their eyes, to improve their vision as they got older. Because the glass discs looked rather like flattened lentils, they were called 'lenses'.

The first lenses were convex, bulging at the centre. Later, opticians created lenses that were thinner in the centre – concave lenses that helped people with short sight. By the year 1450, the ingredients for a telescope existed, but no-one had sussed the recipe for putting them together. There are some tantalizing hints that – well before Galileo – people may have used lenses or curved mirrors to make distant objects seem nearer. The English astronomer Thomas Digges mentioned that his inventor father, Leonard, 'hath by proportional Glasses duly situate in convenient angles, not only discovered things far off, read letters, numbered pieces of money with the very coin and superscription thereof, cast by some of his friends of purpose upon downs in open fields, but also seven miles off declared what hath been done at that instant in private places.'

It sounds like a telescope, but no-one at that time could possibly have made a telescope that revealed inscriptions on coins in distant fields, yet alone things done in private 7 miles (11 km) away. The claims are just too ambitious. Also, it wasn't followed up. When the telescope was actually invented, it was such an amazing and useful device that it spread like wildfire.

In the summer of 1608, a Dutch spectacle-maker called Hans Lipperhey (or Lippershey) found a unique combination of two lenses – one convex and the other concave – could make objects appear much nearer.

If we want a birthday for the telescope, we could take it as September 25,1608. On that date, Lipperhey presented a letter to the States General in The Hague, applying

LEFT *The invention of the astronomical telescope sent shockwaves through Europe, leading to many significant discoveries. Here, Polish astronomer Johannes Hevelius observes the heavens. Over a period of four years, he produced the first detailed maps of the Moon.*

ABOVE *One of the architectural icons of Venice, the 16th-century Rialto Bridge – shown in a detail of a painting here by Canaletto – remains virtually unaltered since Galileo's residence at the University of Padua in the Venetian Republic. In Galileo's time, science, music, art and architecture were at one: in fact, Galileo's father, Vincenzio, was a prominent composer.*

OPPOSITE *Galileo would have been astonished that a spaceprobe with his name on it had an extended rendezvous with the planet Jupiter, starting in 1995. In this artist's impression, Galileo sweeps over volcanic Io – one of the four moons that he observed through his telescopes.*

for a patent for 'a certain device, by means of which all things at a great distance can be seen as if they were nearby.'

Matters quickly got muddy, as two other opticians claimed to have invented the telescope before Lipperhey. But Allan Chapman is in no doubt who should be awarded the laurels. 'Lippershey seems to have been the first person to put a pair of lenses together – certainly he got the recognition by the State General of Holland. He sold them; and within weeks they being used across Europe. We know they were on sale in Paris not very long after. These were novelty devices, really.'

Galileo first got to hear about the new instrument in the summer of 1609, when a vendor arrived from Paris to sell a telescope to the Republic of Venice. Galileo realised that the telescope was not just a novelty, but potentially a military secret weapon. It would let them see enemy ships far out at sea.

While we think of Venice today as a genteel tourist destination, in Galileo's time war was never far from anyone's thoughts. The island city was a major Mediterranean power in its own right. Only 20 years earlier, Venetian warships had led the victorious Christian fleet against the Ottoman Empire at the Battle of Lepanto, near the mouth of the great Gulf of Corinth.

Galileo was then at the University of Padua, the centre of learning for the Republic of Venice. It was the second most important university in Italy, after Bologna, and Galileo had an excellent position as professor of mathematics. He also earned a bit on the side of his academic salary by teaching visiting noblemen the maths of warfare – surveying, architecture and mechanics.

Galileo had hit on a nice little earner that relied on the Republic's penchant for new military inventions. Galileo's 'Geometric and Military Compass' comprised a pair of metal rulers linked by third curved ruler, with engraved scales that allowed the user

to solve just about every mathematical problem of the day, including changing currencies and setting up cannon.

'Galileo was ambitious,' says Allan Chapman. 'I think he genuinely wanted fame. He was 45 years old at the time of the discovery of the telescope, he was working for what he saw as a very unappreciative salary. And all of a sudden, he heard of this little tube, with possibilities. He first milked it financially, not for astronomical reasons, by flogging it to the Serene Republic of Venice.'

With his formidable practical skills, and excellent glass from the Venetian island of Murano, Galileo had soon made a telescope better than any on the market. From the city's bell-tower, Galileo proudly showed the Venetian senators approaching ships so far out at sea that it would be two hours before the regular lookouts could spot them.

And the wily Galileo offered his telescope free. The Doge of Venice gratefully accepted, and doubled Galileo's salary. But he would have to remain in Venice. And Galileo was feeling his home city of Florence beginning to tug at his heartstrings.

Galileo Galilei had been born in 1564, when the great Danish astronomer Tycho Brahe was a student and the mathematician Kepler still to be born. And during his early years the family shuttled between Pisa and Florence. The young Galileo – ironically, given the way things turned out – toyed with the idea of entering the Church. But his father insisted he studied medicine at the University of Pisa.

His father, Vincenzio, was a musician – and a bit of a maverick. In his compositions, he broke the rules by allowing notes that were dissonant. And he experimented with the strings of his lute, measuring how the tension in the string changed the pitch of the note. He commented: 'they who in proof of any assertion rely simply on the weight of authority, without adducing any argument in support of it, act very absurdly.'

It's easy to see how his eldest son (always known by his first name, Galileo) picked up on the importance of practical skills, of testing all his ideas by experiment – and challenging authority.

At Pisa, Galileo found that medicine wasn't his bag at all. He became more and more intrigued by the behaviour of objects in the real world – what we'd now call physics. One day, while attending a service in the cathedral in Pisa, he watched a chandelier swinging from the ceiling. He was intrigued. The lamp seemed to swing back and forth in exactly the same time, no matter how widely it was swinging. It was the first hint of the pendulum clock, the world's first accurate time-keeper. Vincenzio was appalled. A career in medicine would be wonderfully lucrative, he told Galileo, while mathematicians were notoriously poor. This conversation may have had a long-lasting influence on Galileo, because he was forever finding ways to make money from his scientific skills.

Although Galileo left Pisa without his medical degree, he soon made such a name for himself that he was invited back as a lecturer in mathematics. His personality was turning out to be a lot blunter than his colleagues were used to. Galileo saw no point in wearing a gown, and he wrote a 301-line poem excoriating academic dress – to the point of saying it prevented professors from anonymously visiting brothels, so leading their hands into sinful temptation!

OPPOSITE *A demonstration of some gravity: Galileo is alleged to have dropped two balls of different weights (one a cannonball, the other made of wood) from the Leaning Tower of Pisa in front of a packed audience. The balls both hit the ground at the same time.*

ABOVE *The great German astronomer and mathematician Johannes Kepler never met the Italian Galileo, and in temperament they were complete opposites – but together they convinced the world of the truth of Copernicus's theory, that the Earth orbits the Sun.*

Galileo's own hands were busy discovering the laws of physics. According to a famous story, he dropped two balls of different weights from the top of the Leaning Tower of Pisa. To the amazement of the students and learned professors standing below, they hit the ground at the same time.

In fact, this is probably a great scientific myth, as Galileo only reported it very late in life. On the other hand, a Dutch engineer called Simon Stevinus did definitely perform this experiment several years before Galileo's reported escapade, by dropping two lead balls of different weights from a church tower in Delft.

Galileo may not have been the first; but, as so often in life, he was the most thorough. He decided to investigate falling bodies by a very clever trick: by letting balls of various weights roll down a wooden slope.

Opposite Galileo's finger, the Museum for the History of Science in Florence has reconstructed two of his slopes. The curator shows how a metal ball and a wooden ball roll at the same rate. Over the slope, Galileo put little gates equipped with bells. As the ball rolls past, it rings the bell. For the first time, Galileo measured how objects accelerate as they fall.

In these experiments, Galileo was trying to break the stranglehold of the ancient Greek philosopher Aristotle who had said that heavier bodies fall faster than lightweight bodies; and – once dropped – they fall at a constant speed.

Galileo's reputation was now in the ascendant. In 1592 he was offered the plum job in Padua, where he would stay for 18 years – according to his later account, the happiest of his life. Undoubtedly, this was in part due to the loving attentions of a young Venetian, Marina Gamba, who became his mistress and bore him three children.

Around the same time, Galileo started to become interested in astronomy – but only to back up his own ideas on physics. The tides had perplexed mankind from the time our remote ancestors set out to sea. Galileo believed he could crack the problem. He thought the tides go up and down because the Earth is rotating under a fixed bulge of water.

According to Aristotle – and the twin astronomical theory of his fellow Greek, Ptolemy – this couldn't happen, because the Earth was stationary at the centre of the Universe.

Like all scholars of his time, Galileo had a copy of Nicolaus Copernicus's book, *De Revolutionibus*, where Copernicus argued that the Earth both spins around, and travels round the Sun once a year. This idea fitted Galileo's theory of the tides to a T.

He was also sent a copy of the most recent theory of the Universe, *Mysterium Cosmographicum*, by the greatest living astronomer, Johannes Kepler, from his home in Graz (now in Austria).

Though they never met, these twin luminaries were to propel astronomy forward over the next 50 years – the fiery Mediterranean physicist and the introverted German mathematician.

Kepler was the most fervent disciple of Copernicus, and he implored Galileo to speak out. 'When the assertion that the Earth moves can no longer be considered something new,' he wrote, 'would it not be much better to pull the wagon to its goal by our joint efforts...Be of good cheer, Galileo, and come out publicly.'

Shortly afterwards, in 1600, Rome witnessed one of its greatest crowd-pullers: a heretic was publicly burnt at the stake. His name was Giordano Bruno. Bruno believed in the theories of Copernicus; and many later books have said his fate acted as a warning to Galileo not to speak out. Actually, nothing could be further from the truth. Bruno had been ordained as a priest, but then proclaimed his own religious ideas. He was an early believer in natural magic. God wasn't a special divine being, but something you found in every tree and every stone. The Universe is infinitely big, and every world out there was populated. The Devil could be pardoned. Jesus wasn't divine, but merely a clever magician.

When things got too hot in Italy, Bruno headed north. He converted to the Protestant church: but they ended up incensed with his ideas and excommunicated

LEFT *Giordano Bruno was an outspoken theologian and philosopher who also supported the Copernican System. But that was the least of his worries. The Church condemned him for disbelieving the virginity of Mary, holding erroneous opinions about Christ, and believing in the plurality of worlds. On February 17 1600, he was burnt – naked – at the stake.*

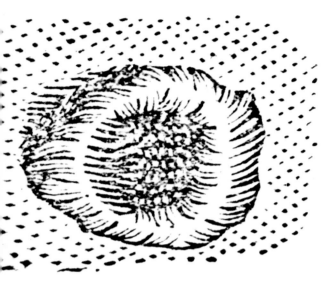

him. His cause wasn't helped by his party trick: using mnemonics to give him a memory so unnatural it seemed he must be in league with the Devil. Back in Italy, Bruno was betrayed to the Inquisition. Refusing to take back any of his ideas, he was condemned as a heretic.

As Allan Chapman puts it, 'saying that Giordano Bruno was burnt at the stake for believing the Earth goes around the Sun, is more or less equivalent to saying that the Boston Strangler got executed because he was a week late taking his library books back!'

When, nine years later, Galileo made his first telescope, he couldn't have imagined that this little device would later bring him face to face with the Inquisition.

In autumn 1609, the ever-curious Galileo turned his telescopes to the sky. First, he looked at the Moon. Aristotle had said that everything in the heavens was perfect and spherical. Now Galileo could clearly see 'the Moon is by no means endowed with a smooth and polished surface, but is rough and uneven and, just as the face of the Earth itself, crowded everywhere with vast prominences, deep chasms and convolutions.'

ABOVE *Apollo 11 astronauts captured the grandeur of this mighty crater – 50 miles (80 km) across – on the far side of the Moon. Like the thousands of other lunar craters, this scar on our satellite's landscape was caused by the impact of a giant meteorite billions of years ago.*

TOP *A sketch of a crater drawn by Galileo, nearly 400 years before this spacecraft image was taken. Ironically, Galileo has only a modest crater on the Moon named after him: it's a mere 9 miles (15 km) across, in Oceanus Procellarum ('the Ocean of Storms'). The crater which*

commemorates Giordano Bruno is not much larger (12 miles/20 km), but, like Bruno's life, it caused waves. Researchers believe that the impact which created the crater was witnessed in 1178 by five English monks – an unprecedented observation of a recent lunar impact.

'Galileo took what was essentially a carnival toy,' says Owen Gingerich, 'and converted it into an astronomical instrument.'

Franco Pacini, Director of the Astrophysical Observatory in Florence, concurs: 'He built a very good telescope for the time. It was the largest in the world. He was a good technologist, and at the same time he was a thinker, because of his previous experiments in physics were all well-thought out, answering definite questions.'

And that marks Galileo out from other people who had turned the telescope to the sky shortly before him. The English scientist Thomas Harriot had peered through his new 'optick tube' at the crescent Moon, while his patron, Sir William Lower, enthusiastically described the Moon as follows: 'In the full she appears like a tart that my cooke made me last weeke; here a vaine of bright stuffe, and there of darke, and so confusedlie all over.'

Galileo wasn't just enthralled with the view. He knew that he was making fundamental new scientific discoveries. These celestial sights were powerful ammunition in his war against Aristotle.

Galileo scanned the glowing band of the Milky Way, and concluded 'all the disputes which have vexed philosophers through so many ages have been resolved. The Galaxy is in fact nothing but congeries of innumerable stars.'

On January 7, 1609, Galileo turned his instrument to the bright planet Jupiter. He was amazed to see three little 'stars' close to the planet. A few nights later, there were four. Night by night, Galileo tracked these objects, and came to the startling conclusion that there were 'four erratic sidereal bodies performing revolutions around Jupiter.'

They had to be moons, circling Jupiter just as our Moon circles the Earth. Other astronomers were merely entranced by the spectacle – like the German astronomer Simon Marius, who later claimed he'd seen these worlds a few days before Galileo.

But Galileo immediately knew the scientific importance of what he'd seen. Many critics of Copernicus's theory said that the Earth couldn't be moving through space, or it would leave the Moon behind. Clearly, Jupiter had some power that made four moons move with it; the Earth could have the same power to hold on to our Moon as the world orbits the Sun.

Galileo rushed into print with his new findings, in *Sidereus Nuncius* ('*Starry Messenger*'), before anyone else could announce them. It was just six weeks from his last observation to publication.

And he had an ulterior motive. Galileo dedicated the short volume to Cosimo de Medici, the Grand Duke of Florence. He named the moons of Jupiter the 'Medicean stars' – though that never stuck, and astronomers now used the classical names proposed by Galileo's rival, Simon Marius (Io, Europa, Ganymede and Callisto).

In addition, Galileo added: 'It was Jupiter, I say, who at your Highness's birth… looked down upon your most fortunate birth from that sublime throne.' Owen Gingerich has also discovered that Galileo drew up a birth chart for Cosimo, sketched on one of his Moon-drawings. Unlike the mystical Kepler, the pragmatic Galileo never believed in astrology – but he had no problem in casting a horoscope when occasion demanded.

BELOW *English astronomer Thomas Harriott drew this pioneering map of the Moon from his observations in 1609-1610. Though Harriott turned a telescope to the Moon before Galileo, he didn't interpret his results in such a revolutionary way.*

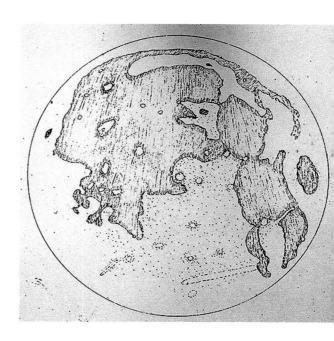

And the ploy worked. Soon Galileo was Mathematician and Philosopher to the Grand Duke of Tuscany. His fame spread around the world. Churchmen, scientists and even poets were spellbound by the Galilean revolution.

In 1611, the English poet John Donne wrote:

And new Philosophy calls all in doubt,

The Element of fire is quite put out;

The Sun is lost, and th'Earth, and no man's wit

Can well direct him where to look for it…

And in these Constellations then arise

New stares, and old doe vanish from our eyes.

And Galileo still had more astronomical discoveries to make. Most important, he saw that Venus can sometimes appear as a crescent, shaped like a narrow Moon. According to the Greeks, this could never happen: the Sun would always light up Venus like a Full Moon. But in Copernicus's system, Venus would appear as a crescent as it swings around on its orbit around the Sun.

BELOW *Detail of Jacopo Tintoretto's interpretation of the origin of the Milky Way: milk gushes from the breast of the goddess Juno as she nurses the boisterous infant Hercules. Because her milk spurted into the sky – and not into the baby's mouth – Hercules missed out on his chance of immortality.*

Galileo also noticed something peculiar about Saturn. When he first looked, there seemed to be a big moon on either side of the planet itself. But later these appendages disappeared; only to come back as two curves. Now we know that he was observing Saturn's rings, though his telescope couldn't make them out clearly. Confused, but determined not to lose priority, Galileo published the letters of the sentence 'I have observed the furthest planet to be triple,' in Latin, as an anagram.

Kepler was one of the recipients. But Galileo's puzzle arrived at a bad time. His wife Barbara had just died, and his patron in Prague, the Emperor Rudolph, had been forced to abdicate. All Kepler could look forward to was a position as schoolmaster in the backwoods town of Linz, in Austria.

Kepler's first project here was essential: to find a new wife. Kepler shortlisted eleven candidates, and then eliminated them one by one. With his normal candour, he noted that one woman had 'stinking breath,' another was 'of insufficient age to run a household,' and in the case of another, 'the contrast of our bodies was most conspicuous: I thin, dried up and meagre; she, short and fat.' Eventually, he chose Susanna, 'with her promise to be modest, thrifty, diligent and to love her stepchildren.'

In comparison with this task, Galileo's anagram might have seemed a doddle! The puzzle-loving Kepler came up with 'Hail, twin companionship, children of Mars' and deduced that Galileo had found two moons orbiting the Red Planet. Amazingly, though Galileo's telescope was too weak to show them, we now know that Mars does possess two small moons.

ABOVE *The real Milky Way. Galileo discovered that it was made of 'innumerable stars,' but nobody followed up his important findings for over two centuries.*

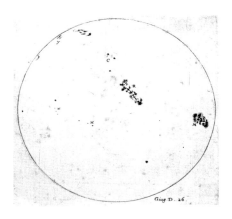

ABOVE *On June 26, 1613, Galileo made this remarkable drawing of spots on the Sun – proving that it was not an unblemished body, as the religious authorities maintained, but had imperfections.*

BELOW *Sunspots, as seen today. These are regions where the Sun's churning gases are suppressed by its powerful magnetic field, and appear dark by contrast.*

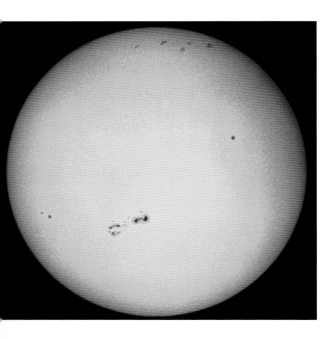

Kepler was no handyman, and couldn't make a telescope of his own. And his eyesight was hardly up to looking through an optick tube. But Kepler did what he did best: he worked out the theory of telescopes. And he came up with a new – and much better – design.

While Galileo's telescope used one convex and one concave lens, Kepler's telescope contained two convex lenses. This gave astronomers a wider view of the sky, and brighter images. The Keplerian telescope has one drawback: everything appears upside down. That meant it wasn't much good for military use, but all subsequent astronomical telescopes based on lenses – refractors – have followed not Galileo's design but Kepler's.

Using a Keplerian telescope, Christopher Scheiner – a Catholic astronomer in Germany – found dark markings crossing the Sun's surface. Galileo, meanwhile, had found sunspots independently. With Galileo's personality, this was bound to lead to a furious argument. It wasn't a good move on Galileo's part, though, as Scheiner was a Jesuit, the order that had the Pope's ear.

Many other churchmen refused to look through a telescope at all; or said they couldn't see anything when they did.

Actually, we have some sympathy for their reaction. In Owen Gingerich's office at the Harvard Observatory, piled up with historic books and papers, he has an exact replica of Galileo's telescope, made by his students. He invites us to take a peek at – well, all there is to see is a brick wall opposite.

And we struggle to make out anything at all. First, the long tube wobbles as you hold it, and every wobble is magnified. Second, it's difficult to bring the telescope into focus, as you try to slide one tube up and down inside the other. And the telescope is literally tunnel-visioned: your view is so restricted that you need real skill to point at anything. We could only admire Galileo's determination in making his groundbreaking observations with such an instrument!

Galileo's new life in Florence was troubled from the start. To keep his two young daughters in safe hands, Galileo put them into a convent. The elder took the name Sister Maria Celeste, and she provided her father with much-needed love and support for the rest of her life.

But his enemies – of whom there were many, thanks to his abrasive manner – began to stalk him. And they had pretty heavy ammunition to use against Galileo. For instance, Joshua told the Sun to stand still; not the Earth. Galileo replied that the words of the Bible are a parable; the writers used simplified language so that complex matters could be understood in everyday terms. He quoted a Vatican librarian, who said 'the Bible is a book about how one goes to Heaven – not how Heaven goes.'

But it was to no avail. In 1616, Galileo visited Rome, to get a definitive ruling. His host was Robert Bellarmine, the Jesuit who had prosecuted and condemned Giordano Bruno.

Bellarmine was a man of science, who had even viewed Jupiter's moons through a telescope, and he was prepared to tolerate Galileo. His main issue was that Galileo

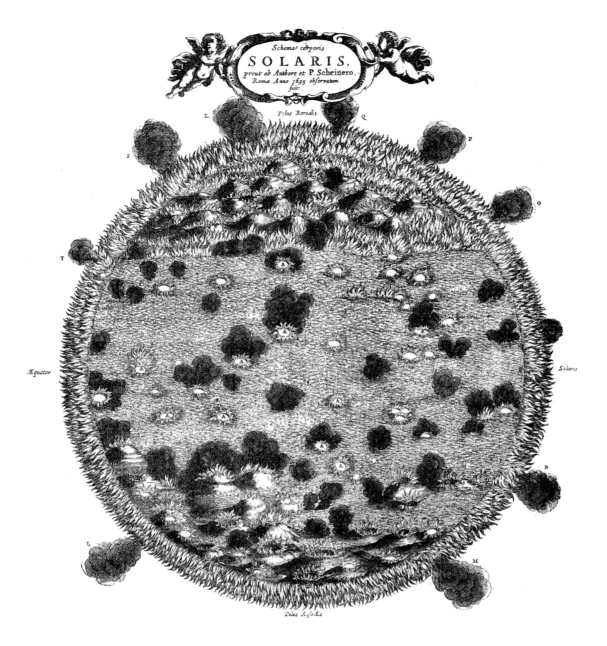

ABOVE *A Victorian engraving of the seventeenth-century concept that the Sun shines because it is a burning mass of coal – an idea that held sway for centuries.*

– as a layman – had no right to make his own interpretation of the Bible. The Catholic Church was becoming very sensitive on this point. The Protestant revolution had carved away huge chunks of its empire, with the rallying cry that people should relate directly to the Bible.

Galileo had certainly stirred up a clerical hornets' nest. Until now, the Church hadn't taken any action on Copernicus's ideas. As long as scientists talked about the moving Earth behind closed doors, everything was OK. But now the Pope convened a special panel of cardinals, who decreed that Copernicus's ideas were 'false and contrary to Holy Scripture.' His book was forthwith banned.

Cardinal Bellarmine exonerated Galileo himself of any wrongdoing, and sent him home with a letter stating that Copernicus's views 'cannot be defended or upheld.'

For several years, Galileo lived peacefully and happily in Florence, pursuing his beloved physics and maths, and keeping clear of astronomy – except in the year 1618, when three comets appeared. Galileo, for once, backed the wrong horse, arguing that comets weren't astronomical objects but merely optical illusions in the atmosphere.

In Austria, Kepler used these comets to push Copernicus's ideas again. He compared comets to living creatures, and suggested that the heavens were as full of comets as the oceans are full of whales.

Kepler was also in the throes of publishing a great new work. *Harmonices Mundi* ('*The Harmony of the World*') was the ultimate expression of the old idea of the Music of the Spheres. With his elliptical orbits, Kepler could work out how a planet's speed changes as it goes round the Sun. He wrote those down in musical notation, with each planet sliding up and down a short scale – at different pitches and at different rates. Mercury has the highest voice, while Saturn is the bass. In recent years, Kepler's *Music of the Spheres* has been recreated digitally. (You can download some versions it on the Web.) The sound is pretty ethereal – it might have pleased the musically adventurous Vincenzio Galileo!

In the midst of this musical mumbo-jumbo, though, Kepler stumbled across a third law that governs planetary motion. There's a simple equation that connects a planet's average distance from the Sun and the period it takes to go around. (Mathematically, the cube of the distance is equal to the square of the period.) That's a rule that astronomers today still find very handy, for instance when they study planets orbiting around other stars.

But Kepler's personal life was far from harmonious. His mother had been arrested, on charges of witchcraft. She was a cantankerous old woman, who had brewed up homemade medicines all her life. And she had made one enemy too many. When Kepler reached her, she was chained up in jail and about to be tortured. She gamely told her captors: 'Even if you were to pull one vein after another out of my body, I would have nothing to admit.' Kepler took on the defense case himself; and eventually arranged for his mother to be quietly released.

RIGHT *Saturn, in its full glory. Galileo's telescopes were insufficiently powerful to resolve the rings, and he saw them as two blobs. The discovery of the nature of the rings would fall to the Dutch astronomer Christiaan Huygens in 1655.*

Back in Linz, Kepler had one task still to complete: the great table of the planets' motions that he had inherited from Tycho. He was still using Tycho's precision measurements, but basing the calculations on his own theory of how the planets move. After 26 years, the Rudolphine Tables were eventually published in 1627.

It had a fantastic frontispiece (see page 126), with a temple containing the 'greats' of astronomy – a Babylonian, the Greeks Hipparchus and Ptolemy, Copernicus and Tycho Brahe. Kepler's great work had been finished just in time; three years later, the Imperial Mathematician died, at the age of 59.

His last work, published after his death, is perhaps the most extraordinary. Somnium tells how an astronomer travels to the Moon, and observes the heavens from this new vantage point. Based on views of the Moon through the telescope, which revealed it as a real world, Somnium is the pioneering example of what we now call science fiction.

Galileo's interpretation of the sky through his telescope, though, had been on the back-burner for a while. He certainly didn't want to stir up any more trouble with the Vatican. But things changed in 1623. An old friend of his, Maffeo Barberini, was installed as Pope Urban VIII. Galileo visited Rome the following year, and the physicist and the Pope had several friendly meetings.

The Italian physicist was sufficiently encouraged to return to an old project of his, the theory of the tides. He set up his book, Dialogue on the Tides, as a debate between a follower of Aristotle and Ptolemy, who argued that the Earth was stationary, and a Copernican scientist who claimed that the Earth orbits the Sun. A third scholar acted as referee.

ABOVE *In this Voyager spaceprobe image, moons Io and Europa float above Jupiter's multicoloured canopy of clouds. The four satellites that Galileo observed orbiting Jupiter convinced him that the Sun, likewise, controlled the planets – and that the Copernican theory was correct.*

Galileo traveled to Rome again with his manuscript, and was told that only a few small changes were needed: the Pope wanted his own views represented, and a different title. Galileo chose *Dialogue Concerning the Two Chief Systems of the World*. Ironically, the new title threw more emphasis on whether Copernicus was right or wrong.

Though Galileo had published previous works in the academic language, Latin, this book was in Italian. Franco Pacini says: 'Galileo advocated scientific communication. He said that people should write in the language of people around, so everyone would know the various steps of scientific research.'

Having checked so carefully beforehand, Galileo was stunned by the Vatican's reaction when the Dialogue was published. Unfortunately for him, the Pope was under extreme political pressure, as the wars between Protestants and Catholics escalated in Europe.

Urban VIII didn't have time to read the book himself, and set up a commission to report to him. It was the opportunity that Galileo's enemies had long been awaiting. They were probably headed up by Christopher Scheiner, the Jesuit astronomer who had crossed swords with Galileo over the discovery of sunspots.

The commission reported that Galileo had overstepped the mark in supporting Copernicus. They also gleefully discovered that Galileo had indeed put the Pope's own words in the book – but in the mouth of the supporter of Ptolemy's system, who was consistently proved wrong. Worse still, Galileo had unflatteringly called this character Simplicio...

When the Pope was told, he 'exploded into great anger,' according to one witness. And he was a man renowned for taking any slight or opposition very seriously.

ABOVE *Katarina, Kepler's mother, stands at the mercy of her captors – accused of witchcraft for creating 'evil brews' (an accusation that was* totally unfounded). *Her accuser is demonstrating the range of torture instruments that would be used to force her to confess. Kepler successfully acted* in her defence – but it delayed the *publication of* Harmonices Mundi, *which also included his third law of planetary motion.*

'It was to do with Galileo's breach of loyalty to his old friend Maffeo Barbarini, now Urban VIII,' says Chapman. 'And he'd made a lot of enemies, he was an abrasive bruiser who had got swollen headed. I think Galileo's love of taking people on basically went too far.'

'I think Galileo overestimated his personal powers of persuasion,' adds Gingerich. 'It wasn't the science so much as the turf-battle as to who had claims to ultimate truth.'

Pacini is more succinct: 'He looked for trouble and he got it!'

The 68-year-old Galileo was summoned to appear in Rome, even though he was in poor health. In April 1633, he appeared before the Inquisition.

The case revolved around his visit to Cardinal Bellarmine, back in 1616. Galileo flourished the affidavit that the Cardinal had given him, where it was clearly written that the system of Copernicus merely 'cannot be defended or upheld.'

The prosecutors had their reply ready. Galileo had been told, they said, that he must 'entirely abandon the opinion that the Sun is the centre of the Universe' and that he must not 'hold, teach or defend it.' Galileo's two-sided Dialogues, didn't defend or hold Copernicus's theory – but it did teach the Sun-centred theory.

OPPOSITE *Originally a friend of Galileo, Pope Urban VIII was renowned for having a short fuse. When he heard about Galileo's book concerning the two systems of the World – which supported Copernicus – he summoned the scientist to trial.*

BELOW *Galileo's villa at Arcetri, near Florence, where he spent the last years of his life under house arrest. But he put his days to good use, studying the problems of mechanics. His efforts were to inspire the work of Isaac Newton, who was born in 1642 – coincidentally, the year that Galileo died.*

Galileo now had little choice in the matter. He had to kneel and read an admission of his guilt, swearing: 'to abandon the false opinion that the Sun is the centre of the world and immovable, and that the Earth is not the centre of the same...'

Some authors have suggested that as Galileo got up he muttered 'Eppur si muove' ('But still it moves'). Who can say? But it's not likely. For all that Galileo was a committed scientist, with a famously brusque manner, he was – more than that – a committed Catholic.

After his recantation, Galileo's enemies were placated, and his friends got his sentence of imprisonment commuted to house arrest; first with the Archbishop of Sienna, and then at Galileo's own villa at Arcetri, near Florence, where he was just a short walk from his beloved daughter, Sister Maria Celeste. Franco Pacini and his colleagues are lovingly restoring this villa to how Galileo would have known it.

Galileo spent his last years carefully avoiding astronomy. Though he couldn't appear in public, he was – says Chapman – 'the world's first celebrity scientist. He became the first scientist to be thought of like Michelangelo or Alexander the Great.'

His final book, *Discourses and Mathematical Demonstrations Relating to Two New Sciences*, concerned the strength of materials and the way that objects move. Because the Vatican had banned any publication by Galileo, he sent the manuscript to Protestant Northern Europe, where it was published in the Netherlands.

BELOW *In this painting by Vincenzo Cesare Cantagalli, Galileo dictates his memoirs. Surrounded by astronomical globes, instruments and clocks, Galileo must have mused on so much. But he looks relaxed; even his faithful hound is in repose.*

Albert Einstein regarded this book as one of the most important in the history of science: 'All knowledge of reality starts from experience and ends in it. Because Galileo saw this, and particularly because he drummed it into the scientific world, he is the father of modern physics – indeed, of modern science altogether.'

Galileo died in January 1642. The Church refused him a celebrity burial. But almost a century later, his body was exhumed and moved to an elegant tomb in Florence's church of Santa Croce. Those present were astonished to find two coffins buried together: the second contained the remains of Galileo's daughter, Sister Maria Celeste. Before Galileo was laid to rest again, his fans removed a vertebra, a tooth and three fingers – including the relic now in the Museum for the History of Science.

By then, the great scientist had become a cult figure. More important, the scientific fire that Galileo had kindled in the cloistered world of Italy had sparked off a blaze of astronomical research throughout Europe. By the time Galileo was reburied, the world had accepted his vision of science and the Universe.

ABOVE *Galileo's inquisition in 1633: in front of the church authorities, the scientist was forced to recant his 'false opinion' that the Earth moved. Only in 1992 did the Catholic Church vindicate his findings, when Pope John Paul II pronounced: 'Thanks to his intuition as a brilliant physicist ... he understood why only the Sun could function as the centre of the World'.*

Matters of some gravity

OPPOSITE *Halley's Comet – seen here in 1986, next to the distant galaxy Centaurus A (bottom right) – was the first celestial object to be ensnared by Newton's law of gravity. Two years before the loss of HMS* Association*, Edmond Halley predicted that this comet would return every 76 years.*

BELOW *The wreck of Britain's flagship HMS* Association *was a matter of considerable gravity to the British Navy. Better astronomical techniques for navigation could have prevented the tragedy.*

The night of October 22, 1707 was foggy and stormy. Through the mist loomed 18 British warships, headed by the *Association*, the great flagship of Admiral Sir Cloudesley Shovell. He was struggling to bring his fleet home for the winter after a campaign in the Mediterranean.

Shovell believed they were just to the west of Brittany: a course to the north-east would take them safely up the middle of the English Channel. But at 7.30 pm, the look-out on the *Association* suddenly saw breakers ahead. Before Shovell could take any action, the ship was on a reef. The Admiral ordered guns to be fired, to warn the rest of the fleet.

But it was too late for the *Association* and three other ships. As they struck the Scilly Isles, the death toll in the maelstrom that night amounted to over 2000 sailors – including Sir Cloudesley Shovell himself.

The British public was incensed. How could the fleet have been so off-course? Some imaginative methods were put forward to help ships find their way on the high seas. For instance, a pair of eminent mathematicians suggested that a line of ships should be anchored along all the world's trade routes, firing mortars in unison.

Loss of H.M.S. Association on the Scillies. 22. Oct. 1707.

By 1714, Parliament had to take some action. Typically, it appointed a panel of experts. This included two astronomers as unlike in their temperament as they were equally revered – the reserved Sir Isaac Newton and the ebullient Edmond Halley.

Newton and Halley were on board because astronomy is the key to navigating around planet Earth. You can find your position north or south of the Equator simply by checking the height of the Pole Star at night, as navigators have known since before the dawn of history. By Shovell's time, navigators also had tables of the Sun's position for any date of the year, and it was easy to measure the Sun's height at midday instead. That sorted out your latitude.

But longitude – your position east or west – was another matter altogether. There's no such thing as a 'pole star' in these directions: everything in the sky seems to wheel around, from east to west, as the Earth spins once every 24 hours.

Time is your best friend here. If it's 12 noon where you are, and you know that it's 6 pm in London, then you must be a quarter of the way round the Earth from London – at the longitude of New Orleans or the Galapagos Islands.

Today it's easy to find out the time in England – you just check it on the Web. But Sir Cloudesley Shovell of course couldn't do that. From the *Association* he could work out his own time; but he still needed to know what the time was in London.

At the meeting, Isaac Newton read out from his carefully prepared notes, comparing four different methods of finding longitude, and sat down. For all Newton's erudition, the committee didn't understand what the great man was actually recommending. After some discussion, though, Parliament offered the Longitude Prize: the staggering sum of £20,000 to anyone who devised a way to measure longitude at sea.

Newton was by this time the elder statesman of British science, and the most famous scientist in the world. As the poet Alexander Pope put it:

Nature and Nature's laws lay hid in night:

God said, 'Let Newton be!' and all was light.

But throughout his life, Newton was almost incapable of relating to other people. It may date back to the traumas of his early childhood. Isaac Newton's father died before he was born, leaving his mother a manor house at Woolsthorpe in Lincolnshire, a reasonable amount of money and a good farm, including a flock of 234 sheep (as compared to an average of 30 to 40). The new baby, born on Christmas morning in 1642, was so small and feeble they could have put him in a quart pot, according to the family legend.

Survive Newton did, but another shock was in store. When he was three, his mother married a neighbouring rector and moved away from home, leaving Isaac in the manor house with his grandmother. Young Newton hated his stepfather, and was never close to his grandmother.

LEFT *Edmond Halley was one of the leading scientists of his age, as an astronomer, geographer, deep-sea diver and pioneer life insurance actuary – as well as being a diplomat and possibly a spy. Unfortunately for him, he worked in the shadow of Isaac Newton, and today is known for little else apart from 'his' comet.*

At school, Newton preferred the company of girls to the boys. He made doll's house furniture for them, and then went on to create intricate working models – of windmills, watermills and even a treadmill turned by a mouse. The daughter of his landlord later recorded a romance between Newton and herself. It may have been simply an infatuation on her part. If real, it was the only time he was ever romantically linked with a woman.

'When you ask about gay astronomers in history,' comments Allan Chapman, 'then Newton is the one who leaps out at you.' In later life, he fell in love with a young Swiss mathematician, Nicolas Fatio de Duiller. The feeling was mutual, though Chapman urges that Newton would have regarded any physical practice with abomination.

When Newton left school at the age of 17, his mother – now widowed – wanted him to run the farm. But this wasn't Newton's forte. In fact, he was fined 'for suffering his swine to trespass in the corn fields.' Fortunately, Newton's uncle and schoolmaster both insisted he went back to classes; and then on to Cambridge. According to one biographer, Newton's servants at Woolsthorpe 'rejoiced at parting with him, declaring he was fit for nothing but the 'Versity.'

Cambridge in the 1660s was dominated by the two great edifices: King's College Chapel and the Great Court of Trinity College – where Newton took up residence.

RIGHT *Dutch scientist Christiaan Huygens was the stepping stone between Galileo and Newton. Huygens invented the pendulum clock (on the wall behind him) and improved the telescope to the point that he was the first to discover the nature of the rings of Saturn. The Dutch physicist's ideas on moving bodies helped to inspire Newton's laws of motion and gravity.*

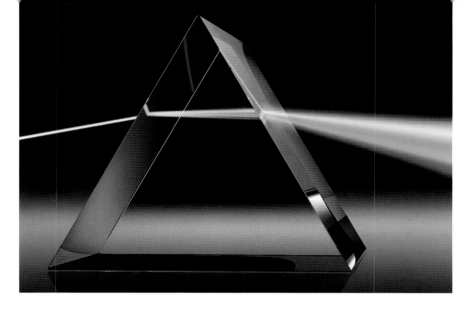

Despite the grandeur of the architecture, Cambridge was academically something of a shambles. The college fellows were not too bothered with teaching; and that pleased the students, as they ended up with degrees without having to work.

Newton was a workaholic, but the system still worked in his favour. He could choose his own curriculum, diving straight into the latest ideas on science and maths. He lapped up Kepler's books, describing how the planets move around the Sun in elliptical paths, and pored over Galileo's measurements of how bodies move.

Galileo had died almost a year before Newton was born. Since then, astronomy had moved on. Nobody now disputed that the Earth moves around the Sun. In France, René Descartes had been trying to understand why: he suggested that space is filled with great vortices, like the swirls in a rushing stream, which carry the planets around.

In the Netherlands, Christiaan Huygens was building ever better telescopes, and discovered that Saturn's two strange 'ears' were in fact a thin ring surrounding the planet. Huygens followed up Galileo's experiments too, and made a pendulum clock – the world's first accurate time-keeper.

Newton was entranced. He'd been taught virtually no science or maths at school; but in a single year at Cambridge he brought himself up to speed with the best minds in the world. He certainly cut a strange figure in college. When Newton became interested in something, he became totally absorbed. He would hardly sleep; and would neglect his meals. As a result of Newton leaving his food on the plate, his cat became notoriously fat.

Newton's curiosity covered everything from perpetual motion to the nature of light. And he wanted to check out everything by experiment. To test how the shape of your eye affects your vision, he stuck a knitting needle behind his eyeball... Aaargh!

But it was maths that first enthralled Newton. In just three years, he single-handedly invented a whole new branch of mathematics, calculus, which is still the backbone for science and technology today. Typically, though, he didn't bother to publish it.

Instead, he moved to investigate the nature of colours. Newton bought a prism, and observed how it projected different colours across his room. Scientists at the time believed that the prism was adding different amounts of darkness to pure white light: bright red was the least tainted, while blue had the most darkness mixed in. Newton's experiments proved something different to him: white light wasn't pure, but was a mixture of literally all the colours of the rainbow.

ABOVE *Woolsthorpe Manor in Lincolnshire was Newton's childhood home – and later the scene of his greatest scientific breakthroughs. A regrowth of the apple tree that inspired his theory of gravity still stands in the garden.*

In 1665, plague struck Cambridge. Newton went home to Woolsthorpe. His mind, of course, was as active as ever. And gravity was now bothering him. According to an early biographer, while Newton 'was musing in a garden it came into his thought that the power of gravity (which brought an apple from the tree to the ground) was not limited to a certain distance from the Earth, but must extend much farther than was usually thought. Why not as high as the Moon, he said to himself.'

An apple tree still stands in the lawn of Woolsthorpe Manor. Is it Newton's tree? For many years, the official answer was 'no': the original tree was blown down in a gale in 1820. But look more carefully, and you'll see that the existing tree grows upwards from the end of a decayed horizontal trunk. In 1998, radiocarbon dating showed that the tree is over 200 years old. So stand in awe when you visit Woolsthorpe Manor: this is indeed Newton's apple tree, standing as new growth from the original fallen trunk and its crop of apples are sisters of those that inspired the theory of gravity.

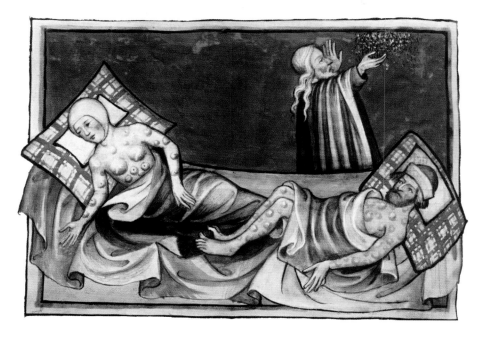

LEFT *Plague struck Cambridge in 1665, leaving its victims suffering from horrendous boils – as seen in this 15th century illustration from the Toggenberg Bible – and a serious risk of death. The closure of the colleges forced Newton back to the seclusion of Woolsthorpe, where he quietly revolutionized science.*

By the age of 25, Newton had laid the foundations for a whole revolution in science. But no-one knew about it. Newton returned to Cambridge after the plague, and carried on his research, closeted away from the rest of the world.

Meanwhile, in London, a group of scientists was trying to answer many of the questions that Newton had already solved. They had started meeting several years earlier at Gresham College, where Christopher Wren was Professor of Astronomy. (Many years later, one of us – Heather – was privileged to hold the same position at Gresham College.) Today, Wren is most famous as the architect who rebuilt London after the Great Fire. But he was also a leading scientist, and his circle of learned friends so impressed King Charles II that he allowed them to call themselves the Royal Society.

The most talented was Robert Hooke, a 30-year-old clergyman's son from the Isle of Wight. According to contemporaries, Hooke was 'of but middling stature, somewhat crooked.' He was always ready to pick an argument, and anyone who disputed his ideas was marked out as an enemy for life. This insecurity marked Hooke's social life as well: rather than marrying a woman of his own social rank, he conducted affairs with his servants, including his teenage housekeeper Grace Hooke – who was also his niece.

'Basically, Hooke was fascinated by instrumentation,' says Allan Chapman. 'He speaks of the telescope, the microscope, the barometer, the air pump – all of which

BELOW *Trinity College, Cambridge, was Newton's domicile during his scientifically productive years. This view, drawn in 1815, looks little different from the college in Newton's time.*

ABOVE *In the 1670s, Johannes Hevelius in Danzig (now Gdansk) built this unwieldy telescope – 150 feet (46 m) long – to minimize the way that its lens distorted his view of stars and planets.*

were new instruments – as being artificial organs that strengthen our native perception and let us see deeper and deeper.'

Hooke's most famous book, *Micrographia*, revealed the miniature world as seen through the microscope – including a drawing of a flea in disgusting detail. He discovered that living matter is made of 'cells' (he coined this word after the cells where monks life in a monastery). Even today, we share our world with many of Hooke's ingenious inventions: the spirit level, the iris diaphragm (used in cameras), the anemometer for measuring wind-speed, the sash-window, and the universal joint (in your car's transmission).

Late in 1671, the Royal Society heard rumours that a mathematician in Cambridge, one Isaac Newton, had invented a new kind of telescope. They borrowed the instrument.

And they were astounded. The secretary wrote back to Newton immediately, informing him that his 'contracting telescope' had been 'examined here by some of the most eminent in Opticall Science and practice, and applauded by them.' The Royal Society sent a description to Huygens, because 'it being too frequent, that new Invention and contrivances are snatched away from their true Authors by pretending bystanders.' They needn't have worried. The great Dutch astronomer added his praise to the 'marvellous telescope of Mr Newton.'

Until then, all telescopes were refractors, using a lens at the front end to gather light and focus it. But the glass in the lens also split the light up into a rainbow, an effect known as chromatic aberration. When you looked at the Moon or a planet, it was ringed by false colours.

The best way out was to give the lens only a shallow curve. And that meant the telescope had to be remarkably long and unwieldy. Poland's leading astronomer, Johannes Hevelius, went to the limit. A wealthy brewer in Danzig (the old town of Gdansk, now beautifully restored), Hevelius married a neighbour who owned two houses and then built an observatory across all three rooftops. To minimize the false colour, his most powerful telescope was 150 feet (46 m) long!

Newton was convinced that a glass lens – like a prism – would always split light into its colours. (For once he was wrong: English optician John Dollond would later make lenses from two different types of glass, so doing away with chromatic aberration.) Instead, Newton made a telescope that had a curved mirror to focus the light.

Before Newton, two scientists had already proposed reflecting telescopes – James Gregory in Scotland and Laurent Cassegrain in France. But neither had the practical

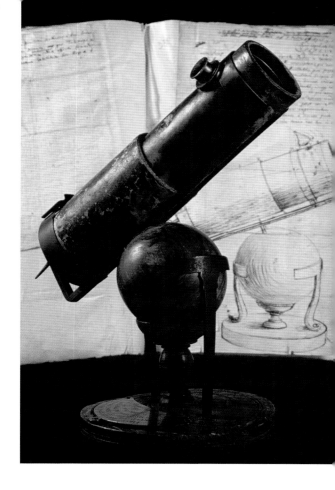

ABOVE *Isaac Newton's revolutionary reflecting telescope used a curved mirror, rather than a lens, to focus light from the stars and planets. It avoided false colours; and was very compact. Here, Newton's own telescope stands in front of his manuscript describing its construction.*

LEFT *In 1991, a total eclipse of the Sun occurred over the world's largest eye on the sky, the Keck Telescope in Hawaii. Like all modern telescopes, the Keck is a reflector, working on the same principle as Newton's original – it just has a mirror 200 times wider!*

skills to build one. Newton took great pride in the fact that he'd constructed the first reflector himself, from casting the metal mirror and carefully polishing it, to fashioning the neat tube. His early experience in doll's house furniture had done him proud.

Today, all large telescopes are reflectors, from the Earth-bound giants on the peak of Hawaii and the Andes of Chile, to the Hubble Space Telescope in orbit.

Chuffed with the response from the Royal Society, Newton sent in a paper describing his experiments on the rainbow spectrum of colours. But Robert Hooke – the Society's resident expert on optical devices – now had his nose firmly put out of joint. He raised a whole raft of irrelevant objections to the theory.

It was the beginning of an enmity that would last until Hooke's death in 1703. Newton wrote a famous letter to Hooke saying 'If I have seen further, it is by standing

on the shoulders of giants.' At first sight, this seems to be a homage to great scientists like Kepler and Galileo; but it's equally likely to be a calculated insult to the hunchbacked Robert Hooke.

The immediate effect, though, was for Newton to hide himself away from the world again, in his ivory tower in the Fens. And the scientists in London soon had their hands full with a regal commission.

The Merry Monarch, Charles II, had recently acquired a French mistress, Louise de Kérouaille, whom he elevated to become the Duchess of Portsmouth. (Through their son, the Duke of Richmond, Louise is a distant ancestor of both wives of the current Prince Charles – Diana Princess of Wales and Camilla Duchess of Cornwall.)

At the time, English people were worried that Louise was spying for her native country. And they were also concerned about the problem of losing ships at sea: the tragedy of Sir Cloudesley Shovell still lay in the future, but Britain's role as the world's naval superpower depended on safely navigating the oceans.

These two matters came together in a quite unexpected way. A Frenchmen close to Louise, an amateur astronomer called St Pierre, whispered to her that he knew how to solve the longitude problem. Louise, in turn, whispered it to the King. In December 1674, Charles II responded by appointing a Royal Commission.

The Commission included Robert Hooke and Christopher Wren; but their secret weapon was a young astronomer named John Flamsteed. Born in Derbyshire, Flamsteed's observations had greatly impressed the Royal Society. He now persuaded the King that England needed a Royal Observatory where astronomers could solve the longitude problem.

That very day, Charles appointed Flamsteed as his 'astronomical observer' – what we'd now call Astronomer Royal. Wren suggested that the observatory be built on the site of a ruined tower, several miles to the east of London, in Greenwich Park.

Charles II wasn't good with backing up his ideas financially, though. The observatory was paid for by the sale of old gunpowder, while the materials were recycled from demolished royal buildings. Flamsteed was unhappy with his salary of £100 per year; and he also had to pay for all his own telescopes and other astronomical equipment.

For all that, Flamsteed had an enviable job. And he worked hard to produce results. What a navigator at sea needed, Flamsteed knew, was a way to look at the sky and deduce what the time was at home. In other words, a celestial clock.

Galileo had suggested using the moons of Jupiter, constantly revolving around the giant planet like four hands of a clock. In principle, it was a great idea. In France, Giovanni Cassini had used Jupiter's moons to measure longitude in a major survey of the country, which reduced its area quite substantially. King Louis XIV was not amused, growling 'my astronomers have lost me more land than my enemies!'

But it was impossible to see Jupiter clearly through a telescope on the rolling seas. Flamsteed needed another timekeeper on the sky. And he had his eye on the Moon.

Every month, the Moon moves right around the sky. You can think of the Moon as the hand of great cosmic clock, moving against the background stars. By carefully measuring the Moon's position, you can tell the time. This was the basis of St Pierre's idea, though it needed a lot more work than he had envisaged.

First of all, the markings on the cosmic dial aren't all carefully spaced like the minute-markings on a clock face: the stars are scattered all over the place. Using his telescopes, Flamsteed would have to measure the positions of the stars far more accurately than the great astronomer of a century before, Tycho Brahe.

Secondly, the Moon isn't a very good time-keeper: it doesn't move at a constant rate across the sky. Flamsteed knew he'd have to measure up the Moon's motion as well. And for navigators to use the Moon, he'd need to predict its position in the future. That was a tough call. At the time, no-one knew why the Moon moved around the sky, so there wasn't a decent theory that would predict its future.

But Flamsteed was determined and obstinate. He spent the next 40 years assiduously observing the heavens on every clear night, even though from boyhood he'd had a weak constitution.

Early on, he received a letter from a student at Oxford. The author, Edmond Halley, was poles apart in personality from Flamsteed. According to David Hughes of Sheffield University – an expert on Halley and his famous comet – 'he always comes over as a very gregarious, happy, hard-working scientist. He was very ambitious, but a very nice chap as well.'

'Halley got – shall I paraphrase – a very good vac job working with Flamsteed,' Hughes continues. After he went up to Oxford, they kept in touch. But there was one difference. Halley's father was wealthy, and gave his son £300 a year to live on: three times the Astronomer Royal's salary. 'So Edmond, as a student at Queen's College, had

RIGHT *Halley's Comet – seen here in a cartoon from 1910 – immortalises Edmond Halley in the public perception. In fact, Halley devoted only a fraction of his life to comet research. He was a rich entrepreneur, whose chief legacy to science was publishing Isaac Newton's ground-breaking new theories.*

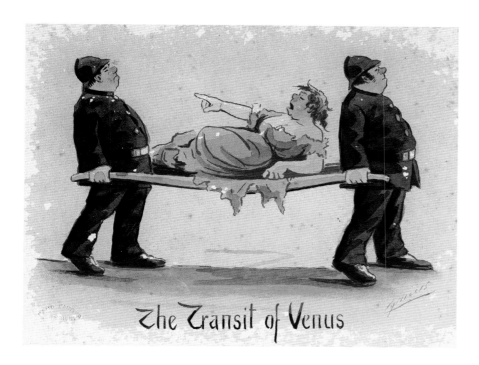

The Transit of Venus

got better instruments that Flamsteed had at Greenwich – it was a very interesting business,' Hughes ruminates.

Halley's father had made his money from soap. 'We're talking the 1660s, when there was the Great Plague of London,' continues Hughes, 'and people suddenly realized that stinking and not washing wasn't a good scheme.' He then invested in houses in London – which unfortunately burnt down in the Great Fire.

Unlike Newton, Edmond Halley didn't want to be closeted up in a college for life. And his father was happy to fund young Edmond's scheme. Halley realised that three talented astronomers were already measuring up the sky that's visible from Europe – Flamsteed at Greenwich, Cassini in Paris and Hevelius in Danzig. So he proposed traveling south of the Equator and charting the little-known skies down there. Otherwise, he feared he would be 'a duck quacking among matchless swans.'

From the British island of St Helena, Halley spent a year peering through clouds to map the stars. When he returned, Flamsteed hailed him as 'our southern Tycho.' He had also observed an unusual daytime event: a transit of Mercury, where that tiny planet passes across the face of the Sun.

Transits of Mercury or Venus are rare phenomena. We were lucky enough to have sunny days in England to view the tiny speck of Mercury crossing the Sun's face in 2003, and the larger black spot of a Venus transit in 2004. But the previous time Venus had passed across the Sun had been when Queen Victoria was on the throne.

The pioneer of Venus transits was a young clergyman called Jeremiah Horrocks. In 1639 – just a few years after Kepler had published his great tables of the planets' motions – Horrocks calculated a transit of Venus would be visible from England. Because it was a Sunday, he had to dash out of church whenever he could to observe the transit.

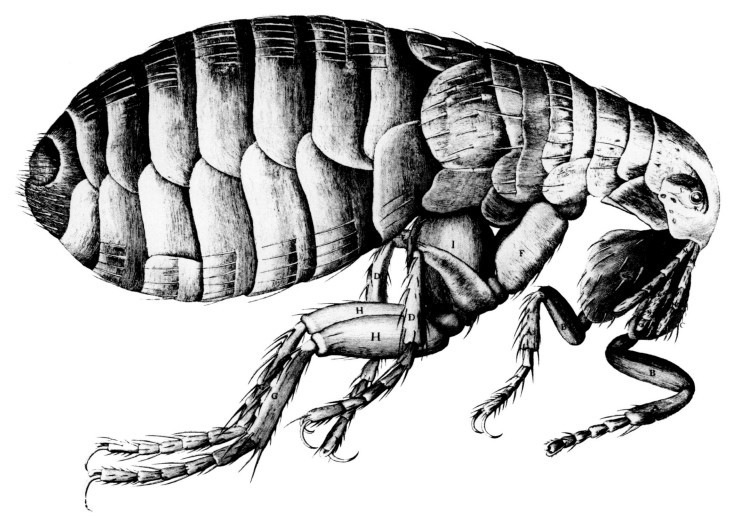

ABOVE *Robert Hooke published this gruesome drawing of a flea in 1665, in an account of his groundbreaking investigations with the newly invented microscope. Hooke was a leading scientist and inventor, but unfortunately, he fell foul of Isaac Newton. After Hooke's death, Newton systematically besmirched his reputation.*

Some 40 years later, Halley realized that transits could provide a way of measuring the size of the Solar System. Venus would give a more accurate result than Mercury, but the next Venus transits wouldn't take place until the 1760s. Halley knew he wouldn't still be about, but he urged his successors to observe them.

And Halley was always politically astute. On his new chart of the southern sky, he designated a group of stars as Robur Carolinum – Charles's Oak – after the tree in which the King had hidden during the English Civil War. Charles II was so pleased, he ordered Oxford University to grant Halley an MA degree, without him having to do any more work!

Halley stayed closely in touch with the King and the Government. 'Later, Halley went on diplomatic missions to the Mediterranean cities to 'look at' the fortifications,' says David Hughes, 'and reading between the lines he was admitted as a scientist, but he was also asked by the British Government to 'spy' on those places to see what they were up to.'

Halley's diplomatic skills were to the fore when the Russian Czar visited London to learn about the latest advances in Europe. After one uproarious dinner at the diarist John Evelyn's house in Deptford, Halley pushed Peter the Great through a hedge in a wheelbarrow!

And he also had to smooth over a row between the irascible Robert Hooke and the Polish astronomer Hevelius. Halley traveled to Danzig to take a look at the rooftop observatory, and reported back that Hevelius's equipment was indeed as accurate as he claimed.

During this time, two comets spread their tails across the heavens: first in 1680, and then in 1682, as Halley was honeymooning in Islington, in north London. The second is the comet that now bears his name.

One evening in January 1684, Halley was drinking coffee with Christopher Wren and Robert Hooke, when the conversation turned to the force that holds the planets in orbit. All three men suspected it was an 'inverse square law': if you move twice as far from the Sun, then the force diminishes to one-quarter.

Hooke bragged that he could prove that this law would make the planets follow elliptical paths. But he wouldn't provide the proof. Then Halley began to hear rumours of a reclusive professor in Cambridge who might have the answer.

Isaac Newton hadn't been idle in the years since he'd lost touch with the Royal Society. But the subject of his research would have shocked his erstwhile colleagues.

First was alchemy. This shows a Jekyll-and-Hyde aspect to the great scientist. He wasn't looking for physical laws in his experiments; but for the mystical side of the arcane art. And – unlike his scientific work – this was not a solitary pursuit. From his notebooks, it seems that Newton worked with other practicing alchemists. Newton's biographer, Richard Westfall, reckons that Newton wrote over one million words on alchemy.

His other passion was theology. Newton believed that the last book of the Bible, the Book of Revelation, was an accurate prediction of the course of human history.

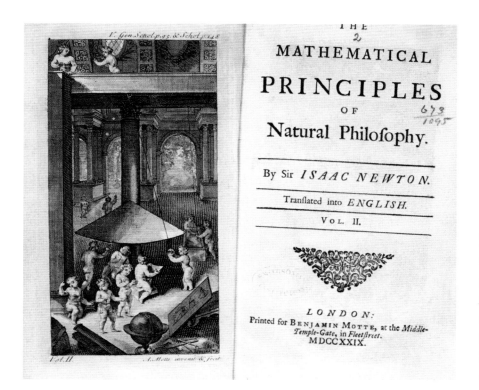

LEFT *Sir Isaac Newton opened a whole new era in science with a three-volume publication that introduced his theory of gravity and the laws of motion to an astonished world. Originally produced in Latin, this is the title page of the first English translation, published some 40 years later in 1729.*

More seriously, he began to doubt whether Jesus was actually an aspect of God. He kept this heretical belief very much to himself. He wasn't in danger of being burnt at the stake; but he would certainly have been thrown out of Cambridge.

'The great fascination with Isaac Newton,' says David Hughes, 'is that he spent a third of his time on biblical chronology, a third of his time on alchemy, and then a bit of his time on physics, maths, optics, the spectrum, gravity, the tides – you name it.'

Halley knew nothing of Newton's other interests when he traveled to Cambridge to meet him. Newton immediately volunteered that he'd already calculated that the path of a planet in thrall to an inverse-square-law was an ellipse. Halley was awestruck. And it hinted that the great mathematician had a lot more to offer. Deploying all his considerable charms, Halley persuaded Newton to write up a longer account.

It was a dizzying moment in the history of science. All his previous thoughts, calculations and proofs seemed to gel in Newton's mind. His thoughts flowed out into three great volumes, under the title *Philosophiae Naturalis Principia Mathematica* – '*The Mathematical Principles of Natural Philosophy*' – now shortened to the *Principia*.

As Newton laboured away, Robert Hooke kept claiming that he had anticipated the results. To a certain extent this was true. Hooke had great insights into science, and he was the first to break free from the ancient idea – dating from the time of Pythagoras – that heavenly bodies naturally move in circles. Instead, Hooke concluded that they 'move forward in a streight line, till they are by some effectual power deflected...'

But it wasn't in Hooke's nature to dig deep. And he didn't have Newton's way with mathematics. Indeed, Newton intentionally made the third book of *Principia* highly mathematical, so that Hooke wouldn't be able to understand it! At the same time, John Flamsteed in Greenwich was very supportive, and provided the observations that Newton needed for the great work.

Halley now found that he had problems in paying for the *Principia*. The Royal Society had lost a lot of money on a previous book, *Historia Piscium* ('*The History of Fishes*') and weren't about to spend a lot more. But Halley stubbornly insisted on the importance of Newton's magnum opus. The outcome was that Halley had to pay for the publication himself. At one stage the Royal Society was so destitute that it paid Halley's wages in copies of *The History of Fishes*!

The Principia basically sets out the underpinnings of how the Universe works. After describing the laws of force and motion, Newton used his law of gravity to calculate the movements of the heavenly bodies. We take it for granted today, but Newton was the first person to state that everything in the Universe exerts gravity on everything else.

Newton's Law of Gravity stood unassailed until 1915, when Albert Einstein introduced his General Theory of Relativity. We need relativity when gravity is immensely strong, as in the vicinity of a black hole. But Newton's gravity works perfectly well in most circumstances. Space agencies like NASA still rely on Newton's law of gravity when they send missions to Mars or Saturn.

And Newton didn't just look at the planets wheeling round the Sun. His theory explained why the Earth's axis is slowly wobbling around, so causing the Precession of the Equinoxes – which gradually shifts the constellations around the sky. It predicted how the Earth should be flattened, like a tangerine, because it's spinning round.

And the *Principia* finally solved the problem of comets – according to Newton, 'this discussion of comets is the most difficult in the whole book.' Until then, many astronomers thought that comets obeyed rules of their own. Newton proved that the comet of 1680 was controlled by the Sun's gravity, like the planets, and that it moved in a very elongated orbit that took it out beyond the orbit of the most distant known planet, Saturn. 'After he'd calculated the orbit of one comet, dear old Isaac said I'm blowed if I'm going to slog my way through a whole gang of others,' Hughes chuckles, 'so he turned to Halley and said – look Ed, I'll give you the observations, just you go away and calculate these orbits for yourself.'

Newton had collected old observations of 24 comets, dating back to 1337. Halley sat down to analyse them: not an easy task, as in those days it took six weeks to calculate the orbit of a single comet.

'And then as he was doing this calculation, he realised that three of the orbits he's calculated were very similar,' says Hughes. 'You can imagine him sitting there and

BELOW *Halley's Comet appears (top left) in the Bayeux Tapestry, which records William the Conqueror's invasion of England in 1066. Here, King Harold of England hears from his advisers that the comet is a harbinger of doom...*

saying: I thought I'd calculated 24 orbits, but I've calculated 22, because one comet's cropped up three times. Whoops, that's periodic then, it comes back every 76 years.'

The celestial visitor we now know as Halley's Comet does indeed put in an appearance once in a lifetime – the last occasion being in 1986, when the European spaceprobe *Giotto* took a close-up look at the cosmic iceberg that forms the comet's core.

Halley had seen this comet on his honeymoon in 1682, and the other appearances on his list were in 1531 and 1507 – when it was observed by Johannes Kepler in Prague and the Welsh astronomer Sir William Lower. Halley now stretched Newton's new theory, by trying to predict exactly when it would next appear. The answer came out to 1758. If he was correct, he wrote 'candid posterity will not refuse to acknowledge that this was first discovered by an Englishman.'

The comet did appear on time, and Halley's fame was assured. But his comet research was only a tiny part of his output – about four per cent, David Hughes calculates. His range was astounding. He invented the first diving bell, to try to raise cannon from shipwrecks. Halley discovered that some of the stars aren't fixed in the sky, but have motions of their own. And he looked into the statistics of deaths, devising the maths behind life assurance.

Halley also persuaded the Navy to put him in charge of a ship in which he sailed the Atlantic measuring the performance of the compass. But Halley wasn't always right. His magnetic measurements convinced him the Earth was hollow, and there could be people living inside.

Meanwhile, Newton had one great challenge left: to explain the precise movements of the Moon as it follows its monthly journey around the Earth. He needed the most accurate observations from Flamsteed at Greenwich. But Flamsteed was getting fed up with the incessant calls from Halley and Newton.

'Flamsteed always thought that the University lads were looking down on him,' says Hughes, 'and of course the biggest of the University gang were Newton and Halley.' With the weight of the Royal Society and Queen Anne behind them, the 'big boys' eventually got hold of Flamsteed's results and published a star catalogue. After the Queen's death, Flamsteed retrieved as many copies as he could, and burned them 'as a sacrifice to heavenly truth.' A few years later, Flamsteed published his own catalogue of 3000 stars, which established Greenwich as the world-leader for measuring star positions.

During this furore, Sir Cloudesley Shovell lost his fleet and his life. At the enquiry, Newton – who had now left Cambridge and was Master of the Mint in London – looked at four alternatives. The suggestion of anchored ships firing mortars was clearly ridiculous. Using Jupiter's moons – as the French had done on land – was impossible at sea. Building a pendulum clock that would work on the ocean swells seemed infeasible. The idea of using the Moon – the original incentive for the Greenwich Observatory – was well under way, with Flamsteed's star catalogue. But no-one yet knew the motion of the Moon well enough.

ABOVE *The Ship and Shovell pub, near Charing Cross station in London, commemorates the unfortunate Sir Cloudesley Shovell, who shipwrecked his fleet on the Scilly Isles – highlighting the practical importance of astronomy for navigating the oceans.*

OPPOSITE *The heart of Halley's Comet was seen in close-up for the first time in 1986, when the European spaceprobe* Giotto *shot through the comet's gas and dust at 150,000 mph (240,000 km/hr). This image shows the comet's nucleus – a ball of ice and dust 9 miles (15 km) long – spewing out steam as it is warmed by the Sun's rays.*

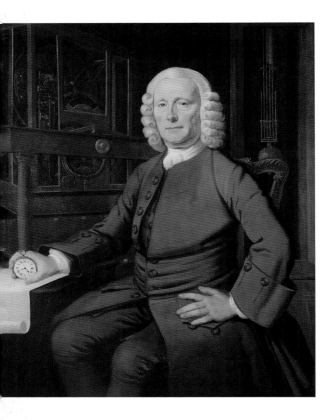

ABOVE *John Harrison was a gifted Yorkshire carpenter, turned clockmaker: his early clock mechanisms were made of wood! Harrison devised the first clocks that would run accurately at sea, and so allow mariners to discover their longitude. Later, he reduced the mechanism to the size of a large watch – seen here in his right hand.*

OPPOSITE *This magnificent clock was the first instrument that could tell time accurately on board a ship. Instead of a pendulum, it has two weights – connected by springs – that oscillate backwards and forwards. It took John Harrison seven years to build this clock, now called H1, and it proved its worth on a test voyage to Lisbon.*

The chance came when Flamsteed died in 1719. Halley was appointed second Astronomer Royal – even though he was then 63 years old. He knew that it would take 18 years to follow all the Moon's complex motions in the sky, but he took to the task with gusto.

In 1730, Halley received a visit from a Yorkshire clockmaker. John Harrison claimed he could build a clock that would keep time at sea. Instead of using a pendulum, his first chronometer – a navigational clock – had two weighted arms that swung back and forth, attached to each other by springs. It was tested on a voyage to Lisbon in 1736. The chronometer performed well, proving that Isaac Newton (by then deceased) had been wrong. But Harrison wanted to do better.

A race – albeit rather tortoise-paced – was now on between the chronometer and the Moon. Halley kept up his observations of the Moon until he died in 1742 – on his favourite chair in the Greenwich observatory, with a glass of red wine to hand. Over in Germany, an astronomer called Tobias Mayer used some new mathematical techniques based on Newton's theory of gravity, along with the Greenwich positions, to publish a table of where the Moon would be over the coming year.

At last, all the ingredients had come together. Sailors could now determine their longitude using the Moon, as Louise de Kérouaille had whispered to King Charles II over 80 years earlier.

But Harrison hadn't been idle. His final chronometer, though, took 30 years to perfect. Looking like an oversized pocket-watch, it performed flawlessly on a voyage to Barbados. Even so, the Government wasn't going to part readily with the prize money. Harrison – then aged 80 – only got the reward after he appealed directly to the King. Apparently George III replied: 'By God! Harrison, I will see you righted!'

By the late eighteenth century, astronomy had nailed down two of the matters that had been of the utmost gravity a hundred years before: how to find our way around our planet; and how the planets kept safely in orbits around the Sun.

The two strands came together in the voyages of the famous explorer Captain James Cook. In his journeyings around the globe, Cook used both methods of navigation, keeping track up the celestial 'clock-hand' of the Moon and watching the tick-tock of a replica of Harrison's chronometer on board his swaying ship.

On Cook's first voyage, from 1768 to 1771, he mapped out the coast of New Zealand, and laid claim to Australia on behalf of the British Crown. Even though he was halfway round the world, Cook could measure his position far better than the unfortunate Sir Cloudesley Shovell just two generations before.

The main reason for his circumnavigation of the Earth, however, involved surveying on the much broader scale. The Royal Society hadn't forgotten Edmond Halley's advice on observing the next transits of Venus. That's why Cook was dispatched on his first voyage.

From a beach in Tahiti, on June 3, 1769, Cook saw the silhouette he'd come so far to observe: 'This day prov'd as favourable to our purpose as we could wish, not a Clowd was to be seen… and the Air was perfectly clear, so that we had every

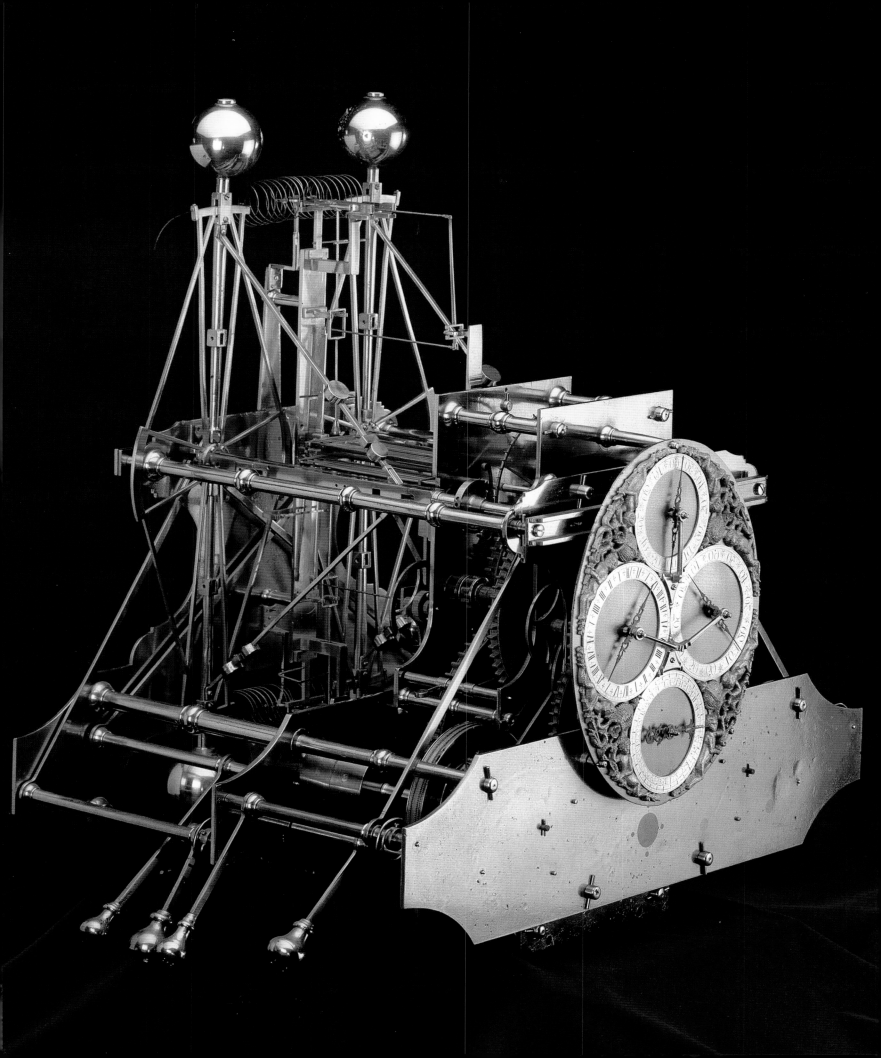

ABOVE *Captain James Cook receives a ceremonial offering from the native Polynesians. His first voyage to the South Pacific was astronomical – to observe a transit of Venus – but he also discovered New Zealand and charted the east coast of Australia.*

advantage we could desire in Observing the whole of the passage of the Planet Venus over the Sun's disk.'

Other astronomers watched the event from Baja California, Hudson Bay and Norway. The French astronomer Jérôme Lalande pooled all their results, and calculated that the Sun lies 95 million miles (153 million km) from the Earth. That's slightly larger than the modern result, 93 million miles (150 million km), but it was the first accurate measurement of something so fundamental in astronomy.

With the distance of the Earth from the Sun now nailed down, Newton's laws revealed that the farthest planet, Saturn, lay a mind-stretching 930 million miles (1.5 billion km) out.

Astronomy had fulfilled its promise from earliest history. It had provided an accurate way of finding our way around the Earth. It had revealed the basic laws of the Universe. And it had given us a true idea of the scale of our planetary system.

The framework of the Solar System as known since antiquity was now in place. And the stage was set for astronomers to explore an unknown Cosmos.

RIGHT *Just as accurate navigation opened up the South Seas to mariners, the magnificent skies of the southern hemisphere provided alluring new horizons to astronomers.*

Planet Hunters

On Tuesday March 13, 1781, a music teacher and composer living in Bath doubled the size of the Solar System overnight. Having despatched his music students earlier that evening, he hurried into the garden of his tall, elegant terraced house in King Street to pursue the real love of his life – astronomy.

'In between 10 and 11, while I was observing the small stars in the neighbourhood of H Geminorum, I perceived one that appeared visibly larger then the rest.'

Little did William Herschel know that he was about to become the first discoverer of a new planet since the days of antiquity.

Herschel was born in 1738 in Germany, the son of a musician in the Hanoverian Foot-Guards. His father Isaac, taught him the violin and the oboe – but, more to the point, instilled in both William and his sister Caroline a deep love of astronomy.

Caroline once reminisced: 'My father was a great admirer of astronomy. I remember his taking me out on a clear frosty night to make me acquainted with several of the most beautiful constellations, after we had been gazing at a comet which was then visible.'

But there were wars to be fought. Young Wilhelm – who would later anglicise his name to William – took up the oboe in the regimental band, but left for England when the Hanoverians were routed by the French at the Battle of Hastenbeck. He became a professional musician, ending up as organist and composer at Halifax, and – later on – at the Octagon Chapel in Bath.

He eventually settled, with his sister Caroline, at 19 New King Street – a three-storied house in an upwardly mobile district of the city. Today, it is a beautifully-restored museum called 'Herschel House', with the gardens looking much as they would have done in the Herschel's time. The Curator, Debbie James, explains why Herschel headed for Bath. 'He came to Bath deliberately, because it was the most fashionable city in the country out of London – and he knew well that if he could get a foothold on the ladder here, he could make it as a musician, because he was very good at what he did.'

'He immediately made an impact. And he had a big fight with Thomas Linley – the Director of Music in the city – which was highly publicised in the Bath Chronicle. Herschel won. Linley moved away, and Herschel took over as Director of Music. Then he was everywhere. He organised concerts all over Bath and Bristol, in the Pump Room, the Assembly Room, in the churches. Everyone knew William Herschel for his musical endeavours.'

But Herschel was typical of many musicians: he was also drawn to mathematics (much as mathematicians are often fascinated by music). He wrote to the mathematician Charles Hutton that 'the theory of music being connected to mathematics has induced me to read all what has been written on the subject of harmony.'

The classic book on harmony at the time was by Dr Robert Smith. Once hooked on the author, Herschel immersed himself in one of Smith's other works: Opticks. Now there was no turning back. Opticks contained descriptions of astronomical telescopes – and Herschel gradually drifted back to his first love.

Herschel, as an impoverished music teacher, couldn't afford to buy telescopes at the time – so he taught himself how to make his own. The easiest type to make was a reflecting telescope – like a giant shaving mirror – using curved light collectors made of metal alloy.

Aided by his devoted sister Caroline – herself a brilliant astronomer, who had left Germany to join him – Herschel cast larger and larger mirrors. On one occasion, the molten metal destined to make a mirror poured out onto the floor. The flagstones cracked, and splinters flew in all directions. Caroline lamented: 'my poor brother fell exhausted by heat and exertion on a heap of brickbats.'

Herschel's obsession was in mapping the sky and attempting to measure distances to the stars. He reasoned that if he could find two stars almost in line, the

OPPOSITE William Herschel – musician, composer, and the first person in history to have discovered a planet. Later in life, he would become the official astronomer to King George III, who set him up with an observatory in Slough, near Windsor Castle. Below is his beloved 40-foot telescope which he constructed there. Servants likened observing with the telescope to 'shaving with a guillotine.'

BELOW The garden of Herschel House, at 19 New King Street, Bath, where Herschel discovered Uranus in 1781.

nearer one should appear to move slightly as the Earth travels around the Sun. So he set himself the task of looking at every reasonably bright star in the sky, to find more distant stars lying in the background.

On that night in 1781, Herschel found a new 'star' in Gemini. With every increase in magnification, the object grew – not what you would expect for a star, which is a point of light. Herschel noted in his journal that he had discovered 'a curious either Nebulous Star or perhaps a Comet.'

The following Saturday night, he recorded that 'I looked for the Comet or Nebulous Star, and found that it is a Comet, for it has changed its place.'

Word of Herschel's discovery started to circulate around the astronomical community. The Astronomer Royal, Nevil Maskelyne, was already speculating that the object could be a planet. And within a few months, astronomers had measured the orbit of the 'Nebulous Star' sufficiently well to reveal that it was not merely a flimsy and fleeting comet – but a new world in the Solar System.

Herschel's finding astonished everyone. He had discovered a world four times larger than the Earth, and almost 15 times heavier. In 1787, Herschel would later go on to discover two of the planet's largest moons (we now know that it has at least 22).

But what to call the new world? Herschel was in favour of 'George' – in honour of the king, George III. However, more established astronomers over-ruled him, being in favour of naming planets after mythological entities. So 'George' became Uranus – a primaeval Greek god representing the sky. He was also the son (and later husband!) of Gaia, the Earth. Between the two of them, they were responsible for creating the family trees of most of the Greek deities.

George III, however, made sure that he didn't miss out on Herschel's discovery. A keen amateur scientist, he called the astronomer to Windsor. There, he set Herschel and his sister up in a house near Slough to serve as the king's private astronomer.

People started to flock from all over the world to meet the dedicated observer, who was now building bigger telescopes than the world had even seen. His greatest was the 'Forty Foot,' which sported a mirror 48 inches (1.2 m) across. Joseph Banks, president of the Royal Society, commented 'my best compliments to Mr. Herschel with wishes that for the sake of science his nights may be as sleepless as he can wish them himself.'

Ably assisted by Caroline, he was delighted to talk to the great and the good about the wonders of the heavens. As Oxford historian Allan Chapman recounts: 'He was a very nice man – the most generous-minded man that you could imagine. He had the knack of warmth and friendship. That's why he got to where he got to at age 50.'

There is an apocryphal story that Joseph Haydn, the renowned composer – when in Britain to conduct one of his London Symphonies – made a pilgrimage to Slough to meet Herschel and look through one of the great man's telescopes. What he saw astounded him: it is rumoured that it inspired him to compose his mighty oratorio The Creation. We ask Allan Chapman if this could be true. 'I can quite believe it,' he replies. 'Haydn was tarred by much the same brush as Herschel – gentle, kind, and with no great enemies.'

ABOVE *Caroline Herschel, William's sister, was his devoted assistant. She even fed her brother by the mouth as he worked on his numerous mirrors. But she was also a great astronomer in her own right, and the discoverer of numerous 'deep-sky objects' – nebulae and galaxies – plus eight comets.*

OPPOSITE *Seven feet (2 m) long, and boasting an alloy mirror 6 inches (15 cm) across, this telescope is similar to the instrument that Herschel used to discover Uranus (and two of its 27 moons). He gave Caroline a comparable telescope, thereby starting her astronomical career.*

BELOW *Bust of William and Caroline in the garden of Herschel House, Bath. Together, the pair changed the course of astronomy: at a time when astronomers were preoccupied with measuring the positions of stars, the Herschels investigated the nature of the Universe.*

And what about Caroline? Little is known, because she destroyed her journals – which recorded her innermost feelings – after her beloved brother married Mary Pitt, of nearby Upton. One can imagine her desolation: for 16 years she had kept house for him, tended the garden in Bath, hand-fed him as he made his monstrous mirrors, and assisted him with his observing.

Fortunately, her legacy lives on. Maybe inspired by the sight of the comet in the sky with her father, she went on to discover eight comets in her own right: the first woman to achieve such a feat. In 1828, the Royal Astronomical Society in Britain awarded her its prestigious gold medal. It would not be until 1996 that the Society bequeathed that honour on a woman again.

William and Caroline Herschel were so much more than stamp collectors of celestial beauties and mysteries. In effect, they changed the course of astronomy. Until the late eighteenth century, it had been dominated by the legacy of Isaac Newton – the emphasis had been in proving that his theory of gravity was right, or wrong. Now the Herschels, with their vistas of new worlds, new cosmic denizens, opened up the heavens – clearing the path for astronomers to address the architecture of our Solar System.

Actually, the person who was perceptive enough to realise that there could be more planets out there had made his prediction as long ago as 1596. That was when Johannes Kepler wrote 'Inter Jovem et Martem interposui planetam' – 'between Jupiter and Mars I would put a planet'.

Kepler was led to this conclusion by formulating his laws of planetary motion. They gave astronomers a scale model of the Solar System: the further a planet is located from the Sun, the slower it orbits. And to Kepler, it was obvious that there was a yawning gap in planetary activity between Mars and Jupiter.

One hundred and seventy years later, the German mathematician Johann Daniel Titius derived a formula to describe the planets' distances from the Sun. This was seized upon by Johann Bode, an eminent and prolific astronomer who was the Director of the Berlin Observatory.

The Titius-Bode Law (often known, rather unfairly, as Bode's Law) is surprisingly simple. If we call the Earth's distance from the Sun 10 units, then the six planets of the time – from Mercury to Saturn – had distances of approximately 4, 7, 10, 15, 52 and 95 units. The large gap between Mars (15) and Jupiter (52) begged a question. There should be a world at around 28 units.

Then Herschel dropped the bombshell. He discovered Uranus (whose name Bode proposed), which turned out to be in an orbit of 192 units from the Sun. It was chillingly close to the Bode's Law prediction of 196. The gap between Mars and Jupiter became not just interesting but embarrassing.

Bode was now on the scent. He and his colleagues – who dubbed themselves the 'Celestial Police' – went on the trail of the missing world...

But they were pipped to the post by Guiseppe Piazzi, an Italian astronomer working in Sicily. On January 1, 1801, Piazzi came across a rapidly-moving 'star' at exactly

ABOVE *The great composer Joseph Haydn in his salon, conducting an ensemble. It is widely rumoured that he was inspired to compose* The Creation *— his magnificent oratorio — after looking through one of Herschel's telescopes when on a visit to England. Other sources cite Handel's music as his muse.*

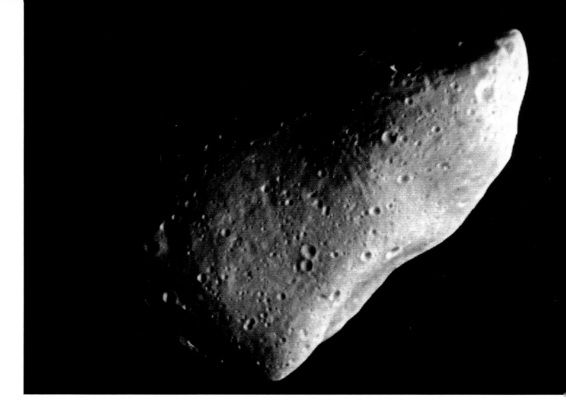

RIGHT *One of hundreds of thousands of 'minor planets' lying between the orbits of Mars and Jupiter, asteroid Gaspra is just 12 miles (20 km) long. It is named after a resort on the Crimean peninsula.*

ABOVE *In the snug bar of a traditional English country pub, author Nigel Henbest receives a citation from the International Astronomical Union telling him that he now has an asteroid named after him. Brian Marsden, the bringer of good news, is Director of the IAU's Minor Planet Center, based in Cambridge, Massachusetts. Asteroid 3795 is now called Nigel – official. And asteroid 3922 is Heather!*

the right distance from the Sun. Piazzi named the new world 'Ceres' – after the patron goddess of Sicily.

William Herschel was clearly not impressed by the discovery. According to Brian Marsden, Director of the International Astronomical Union's Minor Planet Center at Harvard, 'Herschel coined the name 'asteroid' for the object. It was somewhat derogatory. He'd discovered a jolly good planet himself 20 years earlier – and he was looking down on the Sicilian discovery.'

'Herschel's planet was big enough to show a definite disc through the telescope. The new discovery was so small that it appeared like a star – that's what 'asteroid' means.'

Approximately 590 miles (950 km) across, Ceres is only one-third the size of our Moon. And it was shortly to be joined by some even smaller worlds, arrested by the telescopes of the 24-strong Celestial Police. These asteroids – sometimes called minor planets – were clearly in a different league from the likes of Venus or Jupiter.

However, astronomers carried on in the classical tradition of naming objects after ancient heroes and heroines. Pallas, Vesta and Juno followed.

But with increasing telescope power, more and more of these 'celestial vermin' began to appear – and astronomers in the 1920s were becoming increasingly annoyed when they crossed the fields of photographic plates when taking long exposures of the more distant night sky.

Nevertheless, the little beasts had to be catalogued. And while each asteroid is allotted a catalogue number, many have official names – allocated by Brian Marsden's IAU Center.

Now, with over 300,000 asteroids discovered, the classical nomenclature system has run out. Astronomers began to suggest alternatives. Mistresses and pets came onto the agenda. 'It was Mr Spock, named after a pet cat, that caused the trouble,' recalls Marsden. 'Mr Spock came on the list next to a night assistant who had died.'

'It was then decided that pets 'would be discouraged.' But there's one case that's come up in the middle of 2006 – someone wanted to name a minor planet after a budgie that could say 'asteroid.'

BELOW *Many asteroids have minuscule moons in tow. Potato-shaped Ida is more than twice as large as Gaspra, and was the 243rd minor planet to be discovered. Tiny Dactyl is only about 0.6 miles (1 km) across. It is named after mythological beings called the Dactyli, who lived on Mount Ida in Crete – where, it is claimed, the god Zeus was reared in a cave.*

Without appearing to boast, Heather and Nigel are delighted to have a heavenly presence as well. Heather's is asteroid number 3922; Nigel's (awarded more recently) is further up in the pecking order: 3795. 'I did this in the interests of sibling rivalry,' jokes Brian.

So what are these pieces of celestial rubble? Almost certainly, fragments of a tiny planet, weighing-in at just four per cent of the mass of the Moon, that was never able to form. Jupiter's mighty gravity wrecked its chance of ever becoming a world.

The planet-hunters had to set their sights on more distant vistas. And by the 1840s, they had convinced themselves that Uranus was being pulled out of its expected orbit by the gravity of an unknown planet beyond.

Still with the Titius-Bode law in the back of his mind, John Couch Adams – a young mathematician at Cambridge – set down to calculate where the missing planet might

RIGHT *A French engraving shows Urbain Leverrier engrossed in the calculations that led him to believe that there was a major planet beyond Uranus.*

BELOW *The English mathematician, John Couch Adams, came up with calculations that were very similar to those of Leverrier. Eventually – in 1846 – the new planet was discovered at the Berlin Observatory by a German, Johann Galle.*

lie, after his graduation in 1843. By September 1845, he knew where the planet would be found. He informed James Challis, the Professor of Astronomy at Cambridge – of whom many apocryphal stories abound.

Apparently, he was wedded to a small but feisty wife, who was once known to have pulled out a hefty burglar from under their bed – while her husband fled from the room. On another occasion, she was worried about him observing so late. He had got trapped behind the telescope. Fortunately, his plucky wife was able to rescue him.

He did look for the distant world. But on the night that he might have spotted it, it's rumoured that his wife called him in for a cup of tea – and when he returned to the telescope, it had clouded up.

Nevertheless, Challis did give Adams an introduction to the Astronomer Royal at Greenwich, George Airy. It was not to be an auspicious occasion. First of all, they never got around to meeting. Airy was in France when Adams first turned up; and on the second occasion, he was having dinner, and the butler told Adams that the Astronomer Royal could not be disturbed. Adams returned to Cambridge, disappointed.

But Airy did look at Adams's predictions. And the misunderstanding that took place has led to one of the great myths in the history of astronomy: that Airy was arrogant and dismissed them.

'I suppose hindsight is a great teacher,' observes Allan Chapman. 'I've gone through so many of Airy's papers that I'm utterly convinced that he was a profoundly honourable man. I think he didn't really quite know who Adams was.'

'Now, Airy does not seem to have believed that you could discover unknown bodies from the positions of known ones. Of course, Adams had done it theoretically. But there's one thing I can say that Adams never did. At this period there were 30 or 40 men living in Britain and Ireland with whopping great telescopes – they'd have spent every hour they could staring at the sky.'

'Why didn't he put a letter in *The Times* – which he could have done easily – along the lines that the Uranus-disturbing planet was in a certain position – and please have a look for it? All the professional astronomers at that time – like Challis and Airy – were overworked. So Adams should have gone to the leisured, grand amateur community. I just don't know, but he didn't … a letter in *The Times* would have done it.'

Meanwhile, over the English Channel, mathematician Urbain Leverrier at the Paris Observatory was on the same track as Adams. But he was savvy enough to publish his results. On June 1, 1846, he predicted the position of the unknown planet in the sky. But no-one in France wanted to co-operate with him on the search.

His obsession with efficiency made him very unpopular – so much that a contemporary remarked: 'I do not know whether M. Leverrier is actually the most detestable man in France, but I am quite certain that he is the most detested'.

Leverrier set his sights outside his native land. On September 23, he wrote to the Berlin Observatory with his predictions. The observatory's assistant, Johann Galle,

BELOW *New world: Neptune, named after the Roman god of the sea, and imaged here by the* Voyager 2 *spaceprobe in 1989. The gas giant is four times the diameter of the Earth, and has the strongest winds in the Solar System, which blow at over 1200 mph (2000 km/hr). It is intensely blue in colour – possibly as a result of methane in its atmosphere.*

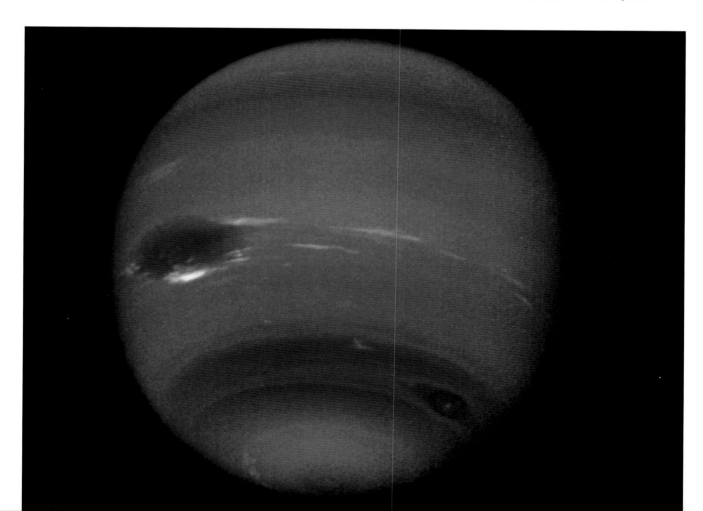

persuaded the director that he should begin a search that very night. Galle was at first disappointed that he couldn't find anything in the region with a clear disc.

Heinrich d'Arrest, a young student present on the observing run, suggested that they should check the stars against a chart, to see if there was an interloper present. Galle called out the positions of stars that he saw: d'Arrest checked them against the chart. Before long, Galle called out a star position that led d'Arrest to exclaim: 'that star is not on the map'. In one night, they had found the eighth planet.

The honour of finding the new planet is now shared between Adams and Leverrier, thanks to the considerable influence of John Herschel – William Herschel's son. He put the young Adams's contributions in perspective. And he arranged for the two men to meet at Collingwood, his residence near Hawkhurst, Kent, in 1847.

A new planet needs a new name. Considerable international argument took place; but in the end, the Berlin observers – who, after all, had discovered the new world – won. Having seen the planet's bluish-green disc for the first time, they opted to name it after the classical god of the sea: Neptune.

It was an appropriate choice. Both Neptune and Uranus are largely made up of water. Almost celestial twins, these gas giants are vast compared to the Earth: Uranus is four times wider than our planet, and Neptune a little smaller. The slightly denser Neptune would outweigh 17 Earths. Both worlds probably boast a rocky core more than five times heavier than our planet. Above this is a huge ocean of hot water, topped with a thick atmosphere of hydrogen and methane. The four gaseous worlds of the outer Solar System – Jupiter, Saturn, Uranus and Neptune – are very different characters from the inner rocky planets like Earth.

But could there be a world beyond Neptune? Maybe an Earth in deep-freeze? A rich Boston banker and amateur astronomer – Percival Lowell – took it upon himself to search for it. He had two obsessions. One was his conviction that there was life on Mars. The other was the belief that there was a planet beyond Neptune. Astronomers had finally come to the conclusion that the discrepancies in the orbit of Uranus could not be down to the influence of Neptune alone.

Lowell built a magnificent observatory in the mountains above Flagstaff, Arizona, and got down to the task of searching for 'Planet X.' He believed that this world was in the constellation of either Gemini or neighbouring Taurus. By now, photography was routine in astronomy, and in the spring of 1915, Lowell was out every night attempting to capture an image of the new world.

Ironically, it did appear on two of his photographs – but it was almost too faint to be seen. Lowell died the next year, unaware that he had actually located his remote world. And the observatory very nearly died as well. Lowell's wife, Constance, contested Lowell's will. Instead of carrying on as an active centre of research, she wanted the observatory to become a museum to her late husband. So obsessed was she with his memory that she carried his clothes with her when she was travelling. And she erected an imposing mausoleum in the grounds – topped with Art Deco tiles – in the shape of the planet Saturn.

BELOW *Rich Boston banker Percival Lowell had two convictions: that there was intelligent life on Mars, and a planet beyond Neptune. He calculated that 'Planet X' was seven times heavier than the Earth, and circled the Sun 50 per cent further out than Neptune. But he failed to recognise Pluto, despite a prolonged search at his purpose-built observatory. In fact, it did appear on two of his photographic plates, but it was much fainter than predicted.*

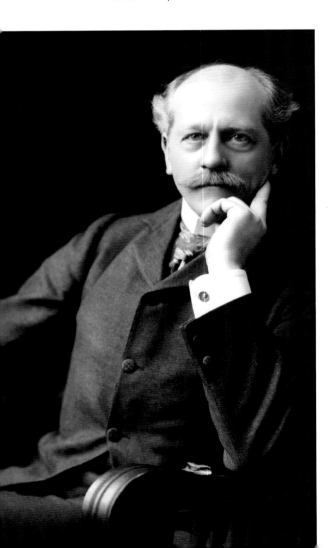

After a 10-year lawsuit, Lowell's nephew gained control of the observatory. He was keen that the observatory should discover 'Uncle Percy's Planet,' and he raised the funds for a new photographic telescope.

The incoming director of the Lowell Observatory, Vesto M. Slipher, decided that he wanted a new member of staff – preferably young and keen – to carry out the interminable process of photographing key areas of the sky night after night. An amateur astronomer from Kansas, Clyde Tombaugh, stepped out of the wings. He had been writing to Slipher with his drawings of Jupiter and Mars, and the director invited him to work with the telescope for a trial period.

Tombaugh soon mastered the intricacies of the telescope, and the blink comparator that he would use to compare the photographs. Tombaugh took two photographs of the same region of sky, and the comparator showed them in rapid succession. He had to set up his photographic plates so that the stars were in the same positions. Then, anything that had moved in the two or three days between the two exposures would jump backwards and forwards, drawing attention to itself.

Slipher first asked Tombaugh to photograph Gemini, where Lowell believed that the planet should lie. But the search failed. Ever an instinctive amateur astronomer, Clyde Tombaugh decided to widen out his hunt to the whole zodiac – the path the planets follow as they criss-cross the sky. Every night, he was at the telescope, photographing the next section of sky; by day, he was at the blink comparator, scanning the plates.

On February 18, 1930, he loaded the blink comparator with plates taken on January 23 and 29. At four o'clock in the afternoon, he suddenly came across a star image which jumped when he blinked from one plate to the other – just the right

ABOVE *In early 1930, 14 years after Lowell's death, young Clyde Tombaugh – recently hired at the observatory in Flagstaff – came up trumps with Planet X. He made a systematic photographic survey of the sky, then compared images of the same region taken a few days apart with his 'blink comparator' (seen here in 1938, when he was searching for even more distant worlds). A moving object would appear to 'jump' when the two plates were viewed in quick succession.*

ABOVE *Over 7000 feet (2000 m) up, amidst the pine forests of Flagstaff, Arizona, the Lowell Observatory is ideally positioned to enjoy a glorious window on the sky. Over a hundred years after it was founded in 1894, the observatory is still at the forefront of modern astronomy – it specialises in looking for asteroids that cross the Earth's orbit, and bodies that lie in the twilight zone at the edge of the Solar System.*

amount to be a planet beyond Neptune. We were privileged to meet Clyde Tombaugh before he died in 1997, and he vividly recounted that moment to us. 'I was terribly excited – I don't think I could ever top that thrill.'

In the most nonchalant manner he could muster, Tombaugh knocked on his Director's office door. 'I announced: Dr Slipher, I've found your Planet X. Slipher rose up, with a tremendous look of elation – and reservation. I found it hard to keep up with him as he went back to the blink comparator'.

For the next few nights, Tombaugh and Slipher checked and re-checked; then went public with the announcement on March 13, 1930. It was exactly 149 years ago to the day since William Herschel had discovered Uranus; and it would have been Percival Lowell's 75th birthday.

In England, the news broke in *The Times* the following morning. Eleven-year-old Venetia Burney was at breakfast with her mother and grandfather at Oxford when the discussion turned to a name for the new world. Venetia had learnt mythology at school, and some astronomy. She suggested that it would be appropriate to name the dim and distant planet Pluto, after the king of the underworld. Her grandfather passed the suggestion to Oxford's Professor of Astronomy, who sent a telegram to Slipher at Flagstaff: 'Naming new planet, please consider Pluto, suggested by small girl Venetia Burney for dark and gloomy planet'.

Lowell had calculated that the new world would prove to be about seven times heavier that the Earth. But as time wore on, Tombaugh and his fellow astronomers began to have their doubts. Pluto appeared to be too small and faint – far too puny to have an effect on the mighty planet Uranus. As time went by, and technology improved, Pluto became the amazing shrinking planet.

Its fate was sealed in the summer of 1978. Jim Christy, of the US Naval Observatory in Washington, was looking at images of Pluto taken near the Lowell Observatory in Arizona. He was puzzled when the images of Pluto turned out to be pear-shaped, unlike the crystal-sharp stars in the background.

Christy's breakthrough was to realise that he was seeing two objects close together – the fainter one possibly being Pluto's moon. He scratches his head and admits: 'I didn't really believe it. And then I later thought: I know it's a moon. I'd better do something about it'. That 'something' was to check other images, and – sure enough – a high magnification revealed that Pluto appeared elongated on other occasions. The elongation of the pear-shape moved around, just as a moon would orbit a planet.

So Pluto's moon came to light, nearly five decades after the discovery of the planet. But what was Christy to call it? He wanted to name it after his wife, Charlene (known as 'Shar') – but fellow-astronomers pointed out that this didn't accord with classical tradition. So he looked in a dictionary of mythology, and discovered, to his amazement, – that Charon – so close in spelling to Charlene – was the ferryman who transported souls across the River Styx to Pluto's underworld. Charon it became.

The name was officially accepted. But all astronomers pronounce it 'Sharon' (instead of the classical 'Karon'), in honour of Charlene. As she tells us: 'Many husbands promise their wives the moon, but my husband got it for me'.

The presence of Charon now allowed researchers to get to grips with the masses and sizes of the two bodies. On the celestial scale, these worlds are diminutive. 1400 miles (2300 km) across, Pluto weighs just one five-hundredth the mass of the Earth. Charon is only one-seventh the mass of Pluto, and is a mere 720 miles (1150 km) in diameter. No way could these bodies have any gravitational influence on distant Uranus. Later – in 1989 – when the *Voyager 2* spacecraft sped past Neptune, astronomers realised that they had hitherto got the mass of the eighth planet slightly wrong. This entirely explained the discrepancies in the orbit of Uranus.

BELOW *Lowell was convinced that discrepancies in the orbit of Uranus were caused by the presence of Pluto. But Pluto is far too small to have any effect on a planet as big as Uranus, which weighs in at 15 times the mass of the Earth. These Voyager 2 spaceprobe images of the distant world – captured in 1986 – show a bland, blue gas-giant (left). A colour-enhanced image (right) reveals hints of cloud-bands. The red region is the planet's pole: Uranus literally circles the Sun on its side.*

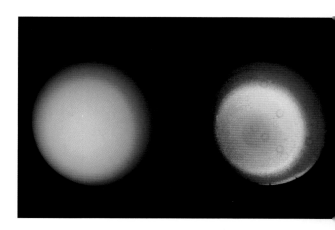

ABOVE *Astronomer Jane Luu, who, with her colleague Dave Jewitt, started the first dedicated search for worlds beyond Pluto.*

BELOW *Warmly wrapped up for the icy conditions atop Hawaii's Mauna Kea, Dave Jewitt – who spearheaded the search for distant worlds – reckons that there are over 70,000 'ice dwarfs' out there with diameters in excess of 60 miles (100 km).*

Even in the 1930s, Clyde Tombaugh wasn't convinced about Pluto's status as a planet. He searched for more massive worlds beyond: but his telescopes weren't powerful enough to turn up any.

It would have to wait until the late 1980s before Tombaugh's baton passed to a modern generation of researchers, with their technology up to the task. Leading the pack was British astronomer Dave Jewitt, who became fascinated with the region of the Solar System beyond Neptune.

'I became interested in astronomy when I was about seven years old – and the only things I could see from London with my tiny telescope were the Sun, Moon and planets. They basically possessed me, and I've never really shaken them off'.

His new search took him far from the light-polluted skies of London to the velvet-black nights of Hawaii. There, with his colleague Jane Luu, he surveyed the heavens with one of the biggest telescopes on the 13,800 foot (4200 m) summit of Mauna Kea.

Their search began in 1987; but it would not to be until 1992 that the first object turned up on their computer screens. 'After five years of searching, it was a surprise to find something', recalls Jewitt. 'It can't possibly be real, we thought. But when it didn't go away – night after night – it became very clear that we had a real object. It was fantastic … many times we'd given up hope that there would be anything beyond Neptune.'

It was to be one of many. But the idea of a swarm of icy bodies at the frontiers of the Solar System was not new. In 1943, the Irish astronomer and engineer Kenneth Edgeworth was one of the first to come up with the idea; Dutch-American astronomer Gerard Kuiper published his theory about a decade later. This swarm is usually referred to today as 'the Kuiper Belt,' unfairly ignoring Edgeworth, and its denizens as Kuiper Belt Objects (KBOs).

By August 1999, the Hawaii team had winkled out 186 of these icy dwarf worlds. And the numbers are still rising: over 800 KBOs are known today. In fact, the amount of material in this region in the Solar System could amount to as much as 30 times the mass of the Earth – far greater than the rocky fragments making up the asteroid belt.

Jewitt and Luu's discoveries were the beginnings of the death-knell for Pluto's status as a planet. 'As far as we can see, Pluto is an absolutely typical KBO', observed Jewitt in 1999, 'but for the fact that it's about three times bigger than any object we've found so far. But that will change very soon – I expect that within the next two or three years, we'll find objects that are as big as Pluto, possibly even bigger.'

How prophetic those words were to be. Planet-hunter Mike Brown from Caltech shares the same territory as Dave Jewitt – scouring the outer reaches of the Solar System. He's fabled for finding more 'biggies' out there than anyone else. But on Wednesday January 5, 2005, he was in for a shock. Staring at his computer screen, he flicked through the results of an observing run at Palomar Mountain on the Samuel Oschin Telescope (which Brown himself had lovingly restored).

© D. van Ravenswaay

His eye was on objects that moved slowly amongst the stars, indicating great distances from the Sun. And there was his latest quarry – but it was surprisingly bright, considering its remoteness. His instinct told him one thing: it had to be bigger than Pluto.

Brown had already made a champagne bet with a colleague that he'd eventually find 'Planet 10' – but he asked for a five-day extension to check the orbit. The observations dated from October 2003, and he needed to confirm all the data on the computer that had arrived over the previous two years.

The Hubble Space Telescope was later to confirm that the object – classified as 2003UB$_{303}$ – was indeed larger than Pluto: but only just. While Pluto's diameter is 1400 miles (2300 km), the new find came in at 1500 miles (2400 km). Its brightness means it must have a very reflective surface – Brown reckons that it must be coated in frozen methane.

Now it came to the vexed question of naming the new object. Brown is renowned for the wacky nicknames he bestows on his new worldlets. Rudolph, Easterbunny and Flying Dutchman come to mind. But on this occasion he surpassed himself, and suggested Xena – the TV heroine of *Xena: Warrior Princess*. He also wanted to name the object's tiny moon Gabrielle, after Xena's sidekick in the show.

But this did not go down well with officialdom. 2003UB$_{303}$ is now known as Eris – in Greek mythology, the goddess of discord. And a very appropriate choice, in the circumstances.

Is it the tenth planet? Rumbles of uncertainty began to stir in the astronomical community. It is so similar to Pluto, which Dave Jewitt rates as a classic KBO. Are

ABOVE *Xena and Gabrielle? Actually, no. Mike Brown, the discoverer of 2003UB$_{303}$ – which is slightly larger than Pluto – wanted to name this object and its moon after TV's Warrior Princess and her sidekick. Convention prevailed: the new world (seen here in an artist's impression) is officially called 'Eris', after the goddess of discord.*

ABOVE *This powerful sculpture of Pluto abducting Persephone was created by Gian Lorenzo Bernini (1598-1680) when he was only 24. Cerberus, the three-headed dog who guards the gates of Hades to prevent souls escaping, looks on. Persephone – the daughter of Earth-goddess Demeter – eventually had to compromise on spending half the year in Hell. When she was in Hades, her sad mother allowed the crops to die; when she returned, the joyous Demeter breathed life into the land.*

these two bodies just part of a huge swarm of icy asteroids living at our Solar System's margins? Do either of them justify the title 'planet'?

Astronomers were faced with a decision. What is a planet? The question had to await a high-level meeting of the world's professional astronomers: the General Assembly of the International Astronomical Union, held every three years.

So in August 2006, we are in Prague to attend the opening ceremony of the IAU. We spend the morning in the bar at the Holiday Inn, next to the conference centre, in the delightful company of the Director of the Minor Planet Center, Brian Marsden. There's to be a vote at the end of the conference as to whether Pluto is a planet – or not.

'Some of us are going to maintain that we should deny both Pluto and UB$_{303}$ major planethood, and go back to the eight planets of 100-150 years ago. Others are going to insist that there are nine planets, and perhaps now 10 – with many more comparably sized bodies to be added in the future. So we would have a stalemate.'

After lunch, we pile into the conference centre, where over 2000 astronomers from all over the world are exchanging greetings. The outstanding Australian radio astronomer Ron Ekers – President of the IAU – mounts the podium.

'At this General Assembly,' he announces, 'we have one of the most serious and interesting debates in the whole of astronomy. The IAU's mandate on nomenclature has to be robust, in well-defined terms. And the definition of a 'planet' has been not well defined for hundreds of years.'

'The IAU has had to be circumspect because of public interest and intense pressure from the media. The discussion has involved not just astronomers …'

'Over the next week and a half, you'll all have your chance for input. At the Closing Ceremony, we'll be taking a vote. After this 26th General Assembly, instead of the Prague Spring, we'll have a Prague Planet Protocol.'

That night, at the drinks reception, we witness – and participate in – the extraordinary scenario of astronomers gathering together in little cliques, talking in whispers about their assessment of the situation.

We can't stay for the final vote. The next morning, we must fly to the beautiful city of Gdansk in Poland, and then drive to the northern town of Frombork, in order to research the legacy of Copernicus.

But it's pretty clear where things are heading.

At the IAU's Closing Ceremony on August 24, 2006, the status of Pluto – and indeed, UB$_{303}$ – is put to the vote. Caltech's Mike Brown is talking to reporters via a teleconference at the time. 'Pluto is dead,' he tells them. The vote also meant UB$_{303}$ (later to be dubbed Eris) is out of the frame as well.

'Pluto is not a planet,' he adds. 'There are finally, officially, eight planets in the Solar System,'

Brian Marsden agrees. 'When it was discovered, Pluto was played up as a planet – hence the minor planet people weren't involved. There are eight major planets – and it should have stayed that way as far as I'm concerned,'

And for fans of Pluto, it got worse. On September 7, 2006, Marsden's Minor Planet Center assigned an asteroid number to the former ninth planet. Now officially called asteroid 134340, it joins Eris (136139) on the list of cosmic castoffs.

Pluto had been completely demoted.

Understandably, some astronomers were not happy. Around 300 of them have refused to recognize the resolution. Most vocal about his grievances is Alan Stern, leader of NASA's New Horizon spaceprobe – which is currently en route for Pluto. He points out that most of the astronomers attending the IAU had gone home by the time of the Closing Ceremony – and that the vote was undertaken by just 424 people. 'I'm embarrassed for astronomy', he despairs. 'Less than five percent of the world's astronomers voted.'

'It won't stand. It's a farce.'

However, Mike Brown – the finder of Eris – is surprisingly sanguine about the decision. 'As of today, I have no longer discovered a planet,' he observes. 'The public is not going to be excited by the fact that Pluto has been kicked out. But, scientifically, it's the right thing to do.'

'For astronomers, this doesn't matter one bit. We'll go out and do exactly what we did. For teaching this is a very interesting moment. I think you can describe much better now by explaining why Pluto was once thought to be a planet and why it isn't now. I'm actually very excited.'

In the end, how does Clyde Tombaugh's family feel about the IAU's decision? His widow, Patricia Tombaugh, muses: 'It's disappointing in a way, and confusing… But I understand science is not something that just sits there.' Her son, Alden, reinforces her point, and looks to both the past and the future. 'This doesn't change my father's achievement. Science is an evolving process… and he was part of that process.'

ABOVE *Earth, Eris, Pluto and its moon Charon – all shown on the same scale.*

BELOW *Mike Brown – no doubt dreaming up more mischievous names for his celestial discoveries!*

Powerhouse of the Stars

Twinkle, twinkle little star
How I wonder what you are.

Throughout the ages, people have been wondering the same thing. The early Greek philosophers had their own ideas about the stars: 'They are compressed portions of air, in the shape of wheels filled with fire, emitting flames at some point from small openings,' claimed Anaximander of Miletus. And as late as the eighteenth century, the English philosopher Thomas Wright of Durham was writing that the stars were giant volcanoes, belching in the darkness.

One thing that all these accounts agree on is that the stars are hot. But that was as far as they could go. Just a century-and-a-half ago, the French positivist philosopher Auguste Comte wrote: 'The field of positive philosophy lies wholly within the limits of our Solar System, the study of the Universe being inaccessible in any positive sense.'

He added: 'We can imagine the possibility of determining the shapes of stars, their distances, their sizes and their movements; whereas there is no means by which we will ever be able to examine their chemical composition, their mineralogical structure, or especially, the nature of organisms that live on their surfaces … Our positive knowledge with respect to the stars is necessarily limited to their observed geometrical and mechanical behaviour.' In his last statement, Comte was partly correct. But he would be proved to be wildly wrong in his assertion that we

RIGHT *Founder of the doctrine of sociology, the nineteenth-century French philosopher Auguste Comte strove to incorporate science into his world vision. His famous statement that we will never be able to ascertain the chemical composition of the stars was understandable at the time – but technological advances would soon kick in.*

would never know the composition of the stars. Already, the seeds of an incredible revolution were being sown …

At the same time as Comte was setting out his thoughts, astronomers were starting to measure distances to the stars. They had been striving to achieve this goal for centuries, but their instruments just weren't powerful enough. The most valiant effort was made by John Flamsteed, England's first Astronomer Royal, who realised that you needed a very long telescope to do the job. Installing himself and his equipment at the bottom of a 90-foot (27 m)-deep well at Greenwich, Flamsteed tried hard. But his 'well telescope' was never successful, and the Astronomer Royal was eventually driven out by the damp.

Although they knew that measuring distances to the stars would be difficult, astronomers acknowledged that the idea behind it was simplicity itself. Look at a star when Earth is in, say, its 'June' position around the Sun; then observe it again in 'December'. If it is nearby, it will appear to 'jump' against the background of more distant stars – an effect called 'parallax'. By measuring the size of the shift, and knowing the diameter of Earth's orbit, it's just a matter of trigonometry to work out the star's distance.

And that's exactly what Friedrich Bessel did in 1838. A German accountant who had taught himself astronomy, his work so much impressed the King that he was given his own, new, observatory at Königsberg. There, he homed in on a faint pair of stars called 61 Cygni. He knew that the duo must be nearby, because of their rapid motion across the sky.

ABOVE *Stars at the tail of Scorpius, the constellation of the scorpion. Red Zeta Scorpii is one of the brightest stars in the Galaxy, with a luminosity of 100,000 suns. Nineteenth-century astronomers – having shaken off the mantle of 'proving Newton right or wrong' when it came to the theory of gravity – were now hot on the trail to find out what the stars were made of, and what caused them to shine.*

Latticed Window
(with the Camera Obscura)
August 1835

When first made, the squares of glass about 200 in number could be counted, with help of a lens.

Over several years, Bessel scrutinised the pair very carefully, using a contemporary telescope built by Fraunhofer (of whom more anon). He succeeded in measuring the tiny parallax shift – and discovered that the star-system lay a staggering 60 million million miles (100 million million km) away.

Unknown to Bessel, the first stellar distance had already been measured when he leapt into print, but the results weren't made known until 1839. The discoverer was a Scottish lawyer-turned-astronomer Thomas Henderson, who was observing half a world away from Bessel, in South Africa. He measured the parallax of one of nearest and most prominent stars in the southern hemisphere – Alpha Centauri.

Meanwhile, Friedrich Struve, at the Pulkova Observatory in Russia, was using Fraunhofer's biggest telescope to concentrate on a beacon in the northern hemisphere – the star Vega. It turned out to be even further away than 61 Cygni – 150 million million miles (240 million million km) away.

But Vega is one of our nearest stellar neighbours. And as time went on, astronomers would go on to measure the distances to much more remote stars.

If it's any consolation, astronomers are no more able to comprehend these enormous numbers than anyone else. So they worked out a shorthand way to express them. Imagine lightbeams, speeding towards you from a star. Light travels at the speed limit of the Universe – 186,000 miles (300,000 km) per second. Astronomers measure the distances to the stars in terms of the time it takes their light to reach us.

For instance, the Sun's light takes 8.3 minutes to reach us, so our local star is 8.3 light minutes away. Alpha Centauri is 4.3 light years away; 61 Cygni 11 light years distant; and Vega lies 26 light years from our Solar System.

Equipped with stellar distances, astronomers were now poised to embark on their continued foray into the cosmos. The next move would come from gentleman-scholar William Henry Fox Talbot – Egyptologist, philosopher, classicist, mathematician, philologist, physicist, botanist – and pioneer photographer. In the late summer of 1835, he made the earliest surviving photographic negative. It was of an oriel window at his ancient home Lacock Abbey, a golden-stone house set in the wilds of the Wiltshire countryside.

Frustrated by not being able to preserve drawings as seen through a camera obscura (an instrument then fashionable to project panoramas of the landscape), Fox Talbot coated the drawing paper with silver salts. When he exposed the image to the Sun, an image miraculously emerged.

It was to be the first intimation of the possibilities of photography. His 1839 paper to the Royal Society was entitled: 'An Account of the Art of Photogenic Drawing or the process by which natural objects may be made to delineate themselves without the aid of the artist's pencil.'

Always painfully shy and modest, Fox Talbot summarised his contribution as follows: 'I do not profess to have perfected an art but to have commenced one, the limits of which it is not possible at present exactly to ascertain. I only claim to have based this art on a secure foundation.'

And he added: 'It will be for a more skilled hand than mine to rear the superstructure.'

That hand belonged to one of Fox Talbot's friends, the astronomer John Herschel, son of planet-discoverer William. 'He was the first person to put photography on a strong footing', maintains Jack Meadows (our former professor). 'The early pictures, you see, faded rapidly with time, and he was the bloke who developed the fixer.'

Soon, astronomers were applying photography to the sky. The Sun, Moon and stars were no longer the victims of fleeting glimpses through a telescope eyepiece: their glory could be recorded in perpetuity. And this set the stage for the next advance in our vision of the heavens – which started, very firmly, on Earth.

Eleven-year-old Joseph von Fraunhofer began his apprenticeship as a glassmaker in Bavaria, under the supervision of a harsh factory owner. In 1801, the building in which they were working collapsed. Thanks to Maximilian IV Joseph, Prince Elector

ABOVE *Pioneer of astronomical photography: John Herschel (son of William), seen here in a classic photographic portrait captured by Julia Margaret Cameron in 1867.*

BELOW *These two early photographs of the Full Moon, taken in 1891, form a stereoscopic pair. Observing one with each eye creates a single three-dimensional image, revealing the ups and downs of the lunar mountains and craters.*

of Bavaria – who co-ordinated the rescue effort – both glassmakers emerged safely from the rubble.

The prince was impressed by young Fraunhofer's skills, and sent him off for formal training. Soon he was producing the finest optical glass in the world, fashioning lenses that would later allow astronomers to measure distances to the stars.

He was also making glass prisms, which split up light into a spectrum (like the bands of colour in a rainbow). On applying a prism to the Sun, Fraunhofer noticed that the spectrum of sunlight was crossed by 574 dark, vertical lines. He was also able to use his prisms on the brightest stars, and noticed that their spectra, too, had dark lines – but not in the same position as those of the Sun.

But Fraunhofer was not a physicist, nor an astronomer. Ever an instrument-maker, he once noted: 'In all my experiments I could, owing to lack of time, pay attention to only those … which appeared to have a bearing on practical optics.'

There had to be an explanation for the dark lines. Fast-forward half a century: the scene still remains in Germany, but to the ancient and beautiful German town of Heidelberg. It lies picturesquely on the River Neckar, dominated by its giant schloss, or castle. Under the brooding shadow of the great pile, two scientists would come together to initiate the next great breakthrough in our knowledge of the Universe.

Enter Robert Bunsen – a daring chemist, and intrepid geologist. He lost his right eye when a compound of arsenic exploded in the lab. And he was sufficiently brave (or foolhardy) to have made measurements of the temperature of the water in Iceland's Great Geyser just before it erupted.

On the customary academic tour of universities early in his career, he befriended the physicist Gustav Kirchhoff. Kirchhoff was only 21 when he formulated the ideas of currents and resistance in electrical networks. Living through a period of political unrest and revolution in Germany, he fortunately sailed through, marrying the daughter of his mathematics professor and raising a family of four children after his wife's untimely death.

The two close friends collaborated together at the University of Heidelberg in the 1850s. They were intrigued by an observation by Fraunhofer: he had used his prisms to conduct experiments on flames. There, he had observed bright lines in the spectrum of the gases, which appeared to be in the same positions as dark lines on the Sun. The duo decided to investigate.

First, there was the apparatus: a more sophisticated version of Fraunhofer's, but still with a prism at its heart. This 'spectroscope' boasted miniature telescopes to make refined measurements. The method was simple: heat a sample of the substances with a Bunsen burner (notorious to millions of schoolchildren from their chemistry lessons), and analyse the light from the gas coming off.

Crucially, the Bunsen burner was non-luminous, so the flame didn't interfere with the gases they were studying. All the gases – which included sodium, lithium, potassium and calcium – emitted a series of bright lines of different colours. And each element

BELOW *Schloss Heidelberg, presiding over the River Neckar. This beautiful German town was home to a revolution in astronomy, when Bunsen and Kirchhoff discovered how to unravel the composition of the stars.*

had a uniquely different pattern of lines. It was as if the team was 'fingerprinting' the elements. Kirchoff then proved that the bright 'emission lines' could be converted into dark 'absorption lines' if you shone a background bright light through the flame.

'It can be concluded that the spectrum of the Sun with its dark lines is just a reversal of the spectrum which the atmosphere of the Sun would show by itself', wrote Bunsen and Kirchoff. 'Therefore, the chemical analysis of the Sun's atmosphere requires only the search for those substances that produce the bright lines that coincide with the dark lines of the solar spectrum.'

It was forensic science applied to the cosmic scene. 'Spectrum analysis – which, as we hope we have shown – offers a wonderfully simple means for discovering the smallest traces of certain elements in terrestrial substances, and also opens to chemical research a hitherto completely closed region extending far beyond the limits of the Earth, and even of the Solar System.'

At last scientists could prove Auguste Comte completely wrong. By identifying the dark lines, astronomers could now determine the composition of stars hundreds of light years away. They were named 'Fraunhofer Lines' in honour of the great glassmaker.

News of Bunsen and Kirchoff's discovery spread rapidly - and in Britain, talented amateur astronomers seized upon it. We ask Allan Chapman if stellar spectroscopy was actually pioneered by amateurs from the Victorian era. 'Utterly and totally,' he replies. 'It's true that the original physics was done in Germany, but it was picked up by people like William Huggins, who lived in Tulse Hill, South London.'

At the age of 30, wealthy young Huggins sold off the family business and settled into an upmarket home in a suburb five miles (8 km) away from the city smog. There he erected his private observatory. In pride of place was a refracting telescope with a lens eight inches (20 cm) across, made by the then master-craftsman of instrument-makers, American Alvan Clark.

Huggins spent his time doing spectroscopy on anything he could point his telescope at. One of his major breakthroughs was to realise that nebulae – 'fuzzy patches' in the sky – differed in their chemical composition, Some were gaseous: stellar nurseries poised to give birth. But others - which we would eventually learn to be galaxies outside our Milky Way - were definitely made of stars.

In his early fifties, he married Margaret Lindsay Murray (later to become Lady Huggins). Much younger than Huggins, she would later go on to become a great astronomer in her own right. She analysed the composition of the Orion Nebula, the great star-forming region, and detected the presence of oxygen. And she was not beyond criticizing her husband's work, as David Hughes from Sheffield University explains: 'It was quite clear that his wife was just as clever as dear old William, and would leave him cryptic notes: 'Dear William: the spectrum you took last night was a bit awful – please try better tonight.'

The Huggins published jointly – their work included the *Atlas of Representative Stellar Spectra* - until age finally overtook them. Jack Meadows muses on their twilight

LEFT *In its final resting place in Sidmouth, Devon – after several adventures around the country – Norman Lockyer's 6¼-inch (16-cm) Cooke telescope peruses the Sun. Lockyer himself described the 1871 instrument as 'second to none.' Today, it is the principal telescope at the Norman Lockyer Observatory for public viewing of the Sun, Moon and planets.*

BELOW *Norman Lockyer (1836-1920) – carrier of Huggins' spectroscopic baton. He was the world's first professor of astrophysics, founder of the scientific journal* Nature, *and discoverer of the element helium in the Sun decades before it was found on Earth.*

years. 'There's a nice story – or perhaps, a semi-tragic story – of them sitting in their old age, hand in hand, as they watched their precious telescope being dismantled because they had become too old to use it.'

But there were other amateurs out there ready and waiting to pick up the Huggins' baton. At Oxford, historian Alan Chapman observes: 'Norman Lockyer comes exactly from this tradition. He's a mixed bag, though, because he becomes an academic astronomer by the back door. He's a civil servant, he gets into astronomy, he tells us that he's become fascinated by doing – you know – good work with a 6 inch [15 cm] refractor.'

'But it's the discovery of spectroscopy that grabs his imagination. And, having done his six-hour-day in the Civil Service, he then becomes Professor at what today we'd call Imperial College in Kensington, London. So he makes that slide into the professional world. Then, in the latter part of his life he leaves it, and becomes a grand amateur astronomer again at Sidmouth.'

It's a beautiful day in early July, and we're in Devon, approaching Sidmouth. The guidebooks are spot-on when they describe the little town as 'a gem of ravishing

Regency architecture.' Lying in the valley of the River Sid, it became a fashionable seaside resort in the 1820s, and the elegant white villas, decorated by intricate ironwork, make it a perfect holiday destination for families who've been coming for generations – including the young Queen Victoria and her parents.

But we're heading for the hills. Above Sidmouth, on Salcombe Hill, is the Norman Lockyer Observatory, established by great spectroscopist in 1911. It boasts several telescope domes, one added by Lockyer's son James: another astronomer hooked on spectra. It's in a stunning location, with walks across endangered heathland, rescued by and cared for by a dedicated team of volunteers – and it's open to the public.

We meet the team who clearly relish looking after Lockyer's legacy. And we talk about Lockyer, the man. Why did he end up in Sidmouth, aged 76? One of the members volunteers that his second wife had relatives in the area. 'He was a workaholic', adds another. 'He had nine children by his first wife, but had a nervous breakdown in his middle years and had to be taken off to recover on a mountain-climbing expedition in Switzerland.'

The team remind us what a remarkable man Lockyer was. Virtually self-taught, and with no formal qualifications, he went on to become the first professor of astrophysics in the world – running the Solar Physics Observatory in Kensington. One of the team, Gerald White, points out that he was also the father of archaeoastronomy. 'He was the

BELOW *Lacock Abbey in Wiltshire, home to gentleman-scholar William Henry Fox Talbot – pioneer of photography. This technique would become the mainstay of astronomical information-gathering until the late twentieth century, when electronic devices eventually took over from the photographic plate.*

first person to sort out the astronomical significance of the Karnak Temples in Egypt. He then went on to look at other megalithic monuments, and in particular, Stonehenge.'

As if all this wasn't enough, Lockyer established the world's first scientific journal which he edited until shortly before his death in 1920. 'Lockyer formed the bridge between the amateur scientist and the professional by becoming the founding editor of *Nature*,' says Jack Wickings, proudly.

Lockyer's passion was the Sun. And in 1868, he discovered an absorption line in the Sun's spectrum that hadn't been detected before. Nothing if not audacious, Lockyer suggested that it had to be produced by a new element, so far unknown to chemists on planet Earth. He named it 'helium', after Helios, the Greek god of the Sun. By the 1890s, astronomers had accepted that a previously unknown element had entered onto the cosmic stage.

Far from being a discipline in which astronomers once meticulously measured positions, or used the stars for navigation, events were now rapidly moving forward into the realms of astrophysics. The Universe was becoming a giant science lab. And at this point, the story shifts to the other side of the Atlantic: the Harvard College Observatory in Cambridge, Massachusetts.

Now it was time to move on from the Huggins' *Atlas of Representative Stellar Spectra* and start to systematise the spectra of the stars. Only by classifying the stars according to the gases that were most prominent in their spectra could astronomers make sense of what the stars were – and how they are born, live and die.

Harvard was home to the most ambitious celestial fingerprinting scheme that had ever taken place. The project was funded with money from the estate of the physician and pioneer astro-photographer Henry Draper, who had a dream of photographing the whole night sky, and compiling a spectroscopic catalogue of the stars. He died of pleurisy aged only 45, his vision unfulfilled – so his widow, Anna Mary Palmer, made a generous donation to Harvard, in order that her husband's wishes could be realised.

The magnum opus eventually emerged early in the twentieth century as the *Henry Draper Catalog* – a mammoth tome containing information on the spectra of over 225,000 stars. Although the catalogue was published under the auspices of the Observatory's Director, E. C. Pickering, the forty-year project was essentially the work of a dedicated team of women 'computers'.

Pickering was well aware that the new generation of educated women with science backgrounds were as competent as men, if not more so. He also felt that the nature of the project – meticulous measurements and calculations – was eminently suited to the detailed scrutiny that only a woman would give it. And he had another reason for hiring women: their wages were less than half that of men, so he could afford to hire large numbers of 'computers' for the enormous survey.

Some of the team – referred to as 'Pickering's Harem' – went on to become highly respected astronomers in their own right. Led by Scots-born Wilhelmina Fleming, who was Pickering's former maid, the ladies included Henrietta Leavitt, Antonia

ABOVE *The man who systematised the stars: Edward Charles Pickering, of the Harvard College Observatory. His groundbreaking spectroscopic classification of our celestial companions – the Henry Draper Catalog – was compiled by a brilliant team of female 'computers'.*

β Aurigæ Dec 1889.

5   10   15   20   25   30

Maury (who was Henry Draper's niece), and Annie Jump Cannon. It's fair to say that Pickering – who was in favour of universal suffrage – kick-started the opportunities for women in astronomy, despite the group's rather disparaging nickname. (Today, half of all astronomy graduate students in the US are female.)

Cannon – who compiled the final version of the catalogue – is responsible for the way in which astronomers classify stars today. Pickering had suggested that stars should be classed in a sequence of decreasing temperature, from 'Type A' to 'Type Q'. By the time Annie Cannon and her team had completed their thorough analysis, the order had changed beyond recognition, running O,B,A,F,G,K,M. Generations of astronomers remember the sequence by the mnemonic 'Oh be a fine girl, kiss me' – although the catalogue's compilers would surely have appreciated it if the word 'guy' had been substituted for 'girl'.

In 1923, Cannon would take a new recruit under her wing. Cecilia Payne-Gaposhkin was born in 1900 in the small English market town of Wendover, amidst the rolling Chiltern Hills (just 6 miles / 10 km away from where this book is being written). She came from a highly intellectual family, whose possessions included the autographs of Charles Darwin and the geologist Charles Lyell.

At the age of five, she spotted her first meteor. The experience affected her in precisely the same way as it touched one of us (Heather). That was it. She was going to become an astronomer. 'I was seized with panic at the thought that everything might be found out before I was old enough to begin', she confessed later in life.

She studied botany, physics and chemistry at Newnham College, Cambridge, where one of her mentors was the great physicist Sir Arthur Eddington. She vividly remembers attending a lecture he gave on Einstein's Theory of Relativity. Later, he would become one of her greatest champions.

While her interest in the biological sciences waned, Payne-Gaposhkin's fascination for astronomy and physics grew. Soon she became an indefatigable cataloguer of stars that changed in brightness.

The meeting of the Royal Astronomical Society in May 1922 was addressed by the new head of the Harvard College Observatory, Harlow Shapley. Young Cecilia was in the audience, and seized on her chance to work at this mecca on the other side of the Atlantic. She wrote to Shapley: 'I am extremely anxious to come … and am prepared to undertake anything that would enable me to work at Harvard …'

Her request came with a stellar list of references, including Eddington – and Payne-Gaposhkin soon found herself heading towards the hallowed ground in Cambridge, Massachussetts. She had imagined that she would continue researches into her beloved variable stars. 'I am extremely glad that you suggest the photographic study of variable stars as a possible line of work for me. I had hoped that you might suggest this subject …'

But under Annie Cannon's expert tutelage, she soon found herself working on spectra. And she would shortly be working on her doctoral thesis. According to astronomical historian Owen Gingerich, quoting the distinguished astrophysicist

Otto Struve: '*Stellar Atmospheres* was undoubtedly the most brilliant Ph.D. thesis ever written in astronomy.' She was the first student – male or female – to earn a doctorate from the Harvard College Observatory.

Gingerich, who arrived at Harvard just over 50 years ago, recalls Payne-Gaposhkin with considerable respect and affection. 'Mrs G – that's what everyone called her. She seemed a formidable, rather remote presence. But she soon struck me as gentle and kind.'

'I enrolled in her 'Introduction to Observational Astrophysics' course. This is where I really learned what 'chain smoking' was. A pack of cigarettes and a single match could get her through the entire period.'

Gingerich is disappointed, though, that she didn't stand by the sensational findings in her thesis research at the time. She had discovered that the composition of the Sun, in the main, was similar to that of the Earth – except in the amounts of hydrogen and helium. Hydrogen, in particular, was vastly more abundant: it appeared to make up almost three-quarters of our local star, and was a million times more abundant than the other elements.

'Many years ago, I asked Mrs G why she had pulled back from what is in retrospect the correct solution for the hydrogen abundances. 'Probably Henry Norris Russell talked me out of it', was her reply.'

Russell was America's leading theoretical astronomer – a man so celebrated with honours and awards that no-one would ever dare to doubt his opinion. He wrote to Payne-Gaposhkin: 'You have some very striking results which appear to me, in general, to be remarkably consistent … there remains one serious discrepancy … it is clearly impossible that hydrogen should be a million times more abundant …'

Sheffield's David Hughes – a huge admirer of Payne-Gaposhkin – leaps to her defence. 'I mean, imagine you're chief Prof. Astronomy Cambridge; you've written lots of books, you're seriously famous, you've got your knighthood. Then you get this spotty research student coming in telling you that, you know, stars aren't made out of earthy material. They're actually made out of hydrogen.'

'You'd just simply say: well, I think you're barmy, my dear.'

'And of course what happens in the history of astronomy is that you get the breakthrough. Then it takes x years, and x can be quite a long time.'

In Payne-Gaposhkin's case, she didn't have long to wait. In 1929, Russell himself produced a monumental 71-page paper in the *Astrophysical Journal* entitled *On the Composition of the Sun's Atmosphere*. While Payne-Gaposhkin had researched the astrophysical evidence, Russell looked into physical nature of the hydrogen atom.

RIGHT *Cosmic Hydrogen Bomb: every second, our voracious Sun converts four million tonnes of matter in its core into energy, which floods out as heat and light. This fusion reaction, which welds hydrogen nuclei into helium, will keep our local star shining for another five billion years.*

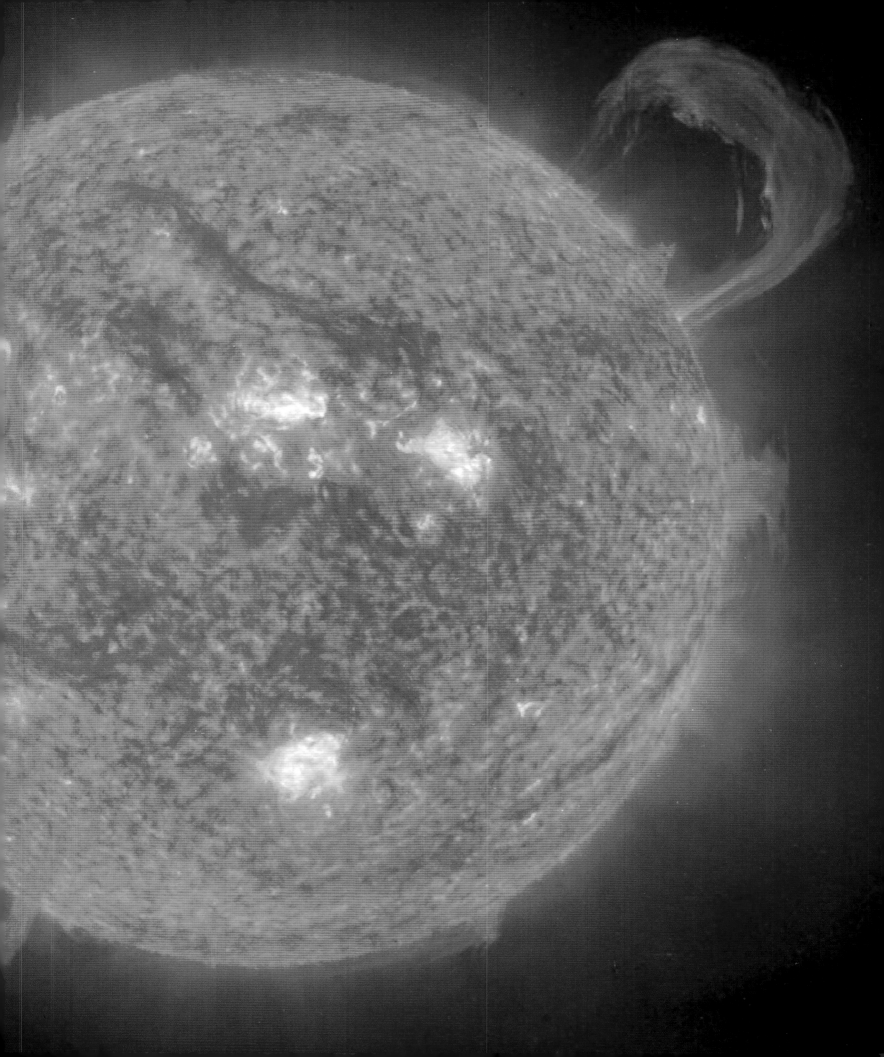

ABOVE *Sun on Earth: a giant mushroom cloud looms over Bikini Atoll in the Pacific after the detonation of an 11-megaton nuclear device – 'Romeo' – in 1954. The nuclear test created widespread radiation damage. One Japanese fisherman died as a result of the explosion, while other crew members suffered long-term illnesses.*

OPPOSITE BELOW *Our local cosmic nuclear reactor – setting in the west – bathes us in life-giving light, heat and energy.*

At the end of his paper, Russell gave full credit to Payne-Gaposhkin's conclusions of 1925 (but said nothing about his original rejection of them). However, he used his considerable influence to convince the astronomical community as quickly as possible that stars are indeed largely made of hydrogen. Payne-Gaposhkin was vindicated at last.

In 1977 – just three years before her death – she received the prestigious Henry Norris Russell Prize from the American Astronomical Society. In her acceptance speech, she told the audience: 'The reward of a young scientist is the emotional thrill of being the first person in the history of the world to see something or to understand something. Nothing can compare with that experience.'

Payne-Gaposhkin had prepared the ground that would now allow astronomers to understand how the stars work.

The clue was hydrogen – and as early as 1927, Sir Arthur Eddington was speculating as to what the powerhouse of a star could be. After Einstein, he was the leading expert on relativity theory, and was well aware that the famous equation $E=mc^2$ (energy = mass multiplied by the speed of light squared) meant that matter and energy were interchangeable.

What might be going on in the core of a star, where pressures were immense and – as Eddington calculated – temperatures would reach millions of degrees? He suggested that stars might generate their energy by nuclear fusion: welding hydrogen into the next element up, helium. Independently, the French Nobel Prize-winner Jean Baptiste Perrin had come to the same conclusion.

But the details of the process would have to wait until 1939, when the brilliant theoretical physicist Hans Albrecht Bethe came up with the mechanism. Bethe left Germany in 1933 after losing his job at the University of Tübingen when the Nazis came to power (his mother was Jewish, although Bethe himself was raised as a Christian). He emigrated to America, where he carved out a dazzling career (earning a Nobel Prize in 1967) until his death at the age of 98 in 2005.

Bethe produced a major paper every decade over the 60-year span of his career. He remembered hearing about a $500 prize for the best unpublished paper about energy production in stars. Bethe won. He recalled: 'I used part of the prize to help my mother emigrate. The Nazis were quite willing to let her out, but they wanted $250 to release her furniture. Part of the prize money went to liberate my mother's furniture.'

Bethe was the senior scientist on the secret Manhattan Project during the Second World War, which eventually led to the development of the hydrogen bomb in America. Later, he would denounce nuclear warfare, arguing that humankind should exploit nuclear power for peaceful means.

Ironically, the Sun is a hydrogen bomb writ large – and the same is true of all stars. But unlike a military hydrogen bomb, the immense weight of the Sun's overlying layers is constantly pressing down on its central nuclear reactor, preventing the energy released from blowing it apart.

As Eddington had predicted, the centre of the Sun is searingly hot – 25 million °F (14 million °C). At these temperatures, atoms can't exist: the electrons that surround the atomic nuclei are moving so quickly that they break away from their dominating protons altogether.

Most of the nuclei are just single protons – hydrogen atoms stripped of their electrons. The high temperatures at the Sun's core make them collide repeatedly, and sometimes the collisions are so forceful that the protons, assisted by uncharged neutrons, are able to overcome their natural electromagnetic repulsion and bond together. The result – two protons and two neutrons – is a helium nucleus.

Luckily for our energy supply, this is a case where the whole is not equal to the sum of the parts. A helium nucleus is only 99.3 per cent as heavy as its four constituents, weighed individually. And so, every time four hydrogen nuclei merge together to make helium, a tiny bit of unnecessary mass is liberated. This mass is converted into pure energy in the heat of the reaction.

The mass-discrepancy seems so small that you might expect only a trickle of energy to escape. But the Sun is vast – and, as a result, it has been able to convert four million tonnes of its matter into energy every second for the past five billion years. Despite this mammoth rate of self-digestion, there's plenty of the Sun left. It has used up less than one-thousandth of its mass by nuclear reactions.

ABOVE *Hans Bethe (1906-2005) was a hugely talented physicist involved in the development of the atomic bomb, but later – with Einstein – he would campaign against it.*

The energy starts life as a blast of gamma rays emerging from the Sun's core. But after travelling a short way upwards, and suffering countless collisions with the Sun's gases en route, the gamma rays lose their initial 'zip.' And there is still a long way to go. By the time the blast of radiation has reached the Sun's surface – some half a million years later – it has lost so much energy that it emerges as heat and light.

Bethe's theory explains what makes the Sun shine – but it doesn't explain everything. Why isn't our local star made exclusively of hydrogen and helium? Its spectrum also contains many absorption lines from heavier elements, such as carbon, sodium and iron. And although hydrogen and helium dominate the spectra of other stars, they, too, reveal traces of other elements. Every star's spectrum is as individual as a fingerprint in its complex pattern of spectral lines. So what causes these differences – and what creates the other elements?

In the mid-1940s, these thoughts were on the mind of a young researcher at Cambridge. Fred Hoyle was particularly puzzled by the large amounts of carbon in the Universe. Carbon, after all, is the atom on which all life is based: without it, we would not exist. But at the time, there was no known way of making carbon. He wrote: 'Would you not say to yourself, 'Some super-calculating intellect must have designed the properties of the carbon atom, otherwise the chance of finding such an atom through the blind forces of nature would be utterly minuscule.'

As an atheist, Hoyle felt that the explanation had to be rational, rather than theological. So he embarked on research that would lead to understanding how other elements are built up in the cores of stars – a process called 'nucleosynthesis'. As his later collaborator, nuclear physicist Willy Fowler, explained: 'The concept of

RIGHT B^2FH- Margaret and Geoffrey Burbidge, Willy Fowler (centre) and Fred Hoyle look on as Willy – on his 60th birthday in July 1971 – gazes adoringly at the miniature steam engine given to him as a present at the Institute of Astronomy, Cambridge. The four were never conventional in their adherence to the old astronomical values. Through their independent approach, they brilliantly devised how the elements we see all around us today were created in stars.

nucleosynthesis in stars was first established by Hoyle in 1946. This provided a way to explain the existence of elements heavier than helium in the Universe, basically by showing that critical elements such as carbon could be generated in stars and then incorporated into other stars and planets when that star 'dies.'

Fred (nobody calls him 'Hoyle') was one of the most innovative minds that the world has ever produced. We were both privileged to get to know this down-to-Earth Yorkshireman. One of us (Heather) was working as a menial research assistant at the Cambridge Observatories, while Fred – next door – was running IOTA: a self-mocking acronym for the Institute of Theoretical Astronomy.

For a young lab assistant, about to embark on her first degree, it was a heady time. Fred would invite his American colleagues (including Willy Fowler) across to attend seminars over the summer. It was a fantastic experience to meet astrophysicists from the international scene.

But Fred was never conventional. He came up with some wildly esoteric theories. The thinly-scattered grains of stardust in space were made of E. coli bacteria, he opined. And comets had brought life to the Earth by delivering it packaged in cosmic eggs. He overstepped the mark several times, especially when he strayed into the territories of other disciplines. John Maddox, the former editor of the prestigious scientific journal *Nature*, remarks: 'the most egregious example was the allegation that the British Museum of Natural History's specimen of *Archaeopteryx* was a forgery.'

'He was the most imaginative of men – a kind of Leonardo,' observes Maddox. Youngsters like ourselves, nurtured on the safe harbour of consensus, saw Fred as someone outside the mainstream. But history will out, and Fred (who died in 2001, aged 86) deserves his exalted place in the cosmic arena.

'He had a lot to say,' recalls Jack Meadows. 'Sometimes he was gloriously right – sometimes he was gloriously wrong.' Jocelyn Bell Burnell, the discoverer of pulsars – and also a Cambridge researcher – remembers Fred's books written for the popular market. 'I wish that more credit was given to them, because they do play an enormous role in bringing people to the subject. But I think Fred was a giant in other ways; an immense intellect.'

Sir Bernard Lovell, founder of the Jodrell Bank radio astronomy observatory, also acknowledges the debt we owe to Fred. 'Well – I think that, unless you have individuals like Fred Hoyle, astronomy settles down into conventional observations. What worries me today is that the young people think that the only things the Universe consists of are the things they're working on. I was a great admirer of Fred Hoyle, and I never believed he was completely wrong.'

When it came to the question as to how stars work, the great meeting of minds took place over several years in Cambridge during the mid-1950s. Geoff Burbidge – an Englishman resident at the University of California at San Diego – was one of the famous quartet who would later reveal the secrets of stellar evolution. Burbidge was there with his wife Margaret, another English émigré, and a leading astronomer in the field of spectroscopy.

LEFT *Wreck of a star: Cas A, which exploded in the constellation of Cassiopeia over 300 years ago. The most powerful radio source in the sky, this supernova remnant is debris from a star whose nuclear reactions went berserk. The image is a composite, captured jointly by the Hubble optical telescope, the Spitzer infrared telescope, and the Chandra X-Ray observatory. Gas at 80°F (25°C) shines red in the Spitzer data; yellow images from Hubble show clouds of gas at 20,000°F (10,000°C); while the Chandra data (green) highlight regions of this cosmic catastrophe glaring at temperatures of 20 million°F (10 million°C).*

'We were both in Cambridge,' Geoff Burbidge tells us, 'when we met Willy Fowler. He was a professor from Caltech, on leave as a Fulbright professor in Cambridge. Willy came to a talk I gave where I was trying to understand some work that Margaret and I had done on the origin of heavy elements in stars.'

'Willy was intrigued by this. And in the process, he introduced us to Fred Hoyle, who he'd known from the time when Fred had been in California. Then we all started working together on the origin of the elements.'

'I mean basically, Fred was the founding genius. Willy was the experimental nuclear physicist, I was someone who could bring a lot to it in terms of creative physics. I was a theoretical physicist and Margaret was the observer.'

Geoff Burbidge, of all astrophysicists, understood Fred more than anyone. 'Well – he was a Yorkshireman. He was very informal, very straightforward. He didn't mind who he talked to or argued with – he was more interested in people than in titles. He was no good on committees, and he hated all the nonsense in Cambridge – the back-biting and all that. But Fred was the most creative person I've ever worked with.'

Over the years, the team worked together at Cambridge and at Caltech in Pasadena, California. And in 1957, they were ready to publish their magnum opus, which – to this day – is always referred to as B^{2}FH (Burbidge, Burbidge, Fowler & Hoyle).

The paper was one of the biggest breakthroughs in our understanding of the Universe. It explained how stars are the crucibles for everything we see around us: seas, mountains, clouds – and life itself.

A star like the Sun – now in the prime of its life – was born almost five billion years ago. It had its origin in a wraith-like cloud of gas and cosmic dust drifting in interstellar space. Then gravity bit in. The cloud began to contract; and the collapse was unstoppable. Slowly but surely, the cloud disintegrated into clumps: protostars.

These were still not proper stars. But there comes a time when a protostar's temperature – which has been rising under its extreme compression – reaches the point of no return. When the core of the protostar hits 20 million°F (10 million°C), it is forced to undergo fusion reactions. Its nuclear fires ignite, energy floods through, and the inpull of gravity is – temporarily, at least – vanquished. A star is born.

As Hans Bethe discovered back in the 1930s, Sun-like stars shine because hydrogen is 'burning' to helium at the core. But eventually – in the case of the Sun, some five billion years in the future – the hydrogen at the centre will run out. The star's core is choked with helium ash.

Fred's masterstroke was to work out what happened next. Bethe himself had written: 'No elements heavier than helium can be built up in ordinary stars.' Fred knew that the answer had to lie in carbon. He predicted a new kind of reaction. When three nuclei of helium come together simultaneously, said Fred, they must stick together and make carbon. This reaction would only work if carbon could vibrate with a particular energy. Fred told his physicist co-conspirator, Willy Fowler, that a carbon nucleus must have this particular 'resonance.' Fowler went to his lab – and lo and behold, Fred was right.

When a star is past its prime, and burns helium in its core, it also suffers serious middle-age spread. It swells up to become a red giant star. In the future, our bloated Sun will expand to swallow up Mercury and Venus. From the Earth, we'd see the distended Sun filling half the sky: the oceans will boil away, and our planet will become a sterile desert. And the smart money now is that the Earth will be cremated as well.

Eventually, the helium fuel at the core will in turn run out. At this critical stage, the Sun will puff off its outer layers into space, to form a glorious cosmic smoke-ring. These objects, many stunningly imaged by the Hubble Space Telescope, are rather confusingly called 'planetary nebulae' – because William Herschel, the first astronomer to study them in detail, thought these round glowing discs looked like a dimmer version of Jupiter or Uranus.

A planetary nebula is chock-full of carbon and other elements that have been forged inside the star – now wafted away into space. The star's core shrinks to become a dying relic, a white dwarf, where the remaining mass of the star is squeezed into a globe no bigger than the Earth.

But in a heavier star, there's another act to be played out. Instead of going down the planetary nebula route, B^2FH found new nuclear reactions, which carry on playing the alchemists' game of transmuting one element into another: carbon into neon, magnesium and silicon. These reactions are hidden in the heart of a supergiant star, which may be thousands of times larger than the Sun.

RIGHT *A phoenix arises from the ashes. The Orion Nebula, 1500 light years away, is the closest region of massive star formation to the Earth. Its material has been garnered from previous supernovae – and now this recycled matter is busily forming into new stars and planets. Our Sun and Solar System may well be the legacy of a supernova explosion.*

Eventually, as B^{2}FH found, the core turns to iron. This poses a severe problem. Iron is the most stable nucleus. If you try to build it up to heavier elements, then the reaction takes in energy. The result? With no further reactions to fuel it, the core suddenly collapses. The matter in the centre is squeezed to its limit – to a ball of neutrons only a few miles across (a neutron star or pulsar) – or even a black hole. The incandescent fury of the core's collapse sends a burst of energy through the star's outer layers, blasting them into space as a supernova.

The gases speeding out from the supernova contain a cocktail of all the elements made inside the heavyweight star – plus other substances created in the blast. This is the origin of the heavy elements we have around us, such as lead, silver and gold.

The crowning achievement of Fred and his colleagues was to show that their theory of the deep inner workings of the stars does predict exactly the proportions of all the elements we find around us in the cosmos.

So – in the answer to 'Twinkle twinkle little star – how I wonder what you are?': we need go no further than look at the world around us. The carbon in our bones, the gold in our wedding ring, the minerals in Earth's rocks … all are made of stardust.

And, Fred – we miss you.

ABOVE *New Zealand amateur astronomers Noel Munford and Ian Cooper captured this spectacular image of Supernova 1987A (lower right), which was seen to explode in the Large Magellanic Cloud over 20 years ago. It was the closest supernova to appear in the skies of Earth since 1604, and its remains will later trigger an orgy of starbirth. The Tarantula Nebula (left) is already a region of prolific star formation.*

The Universe beckons

The Universe was created at nightfall immediately preceding Sunday, October 23, 4004 BC. Well: that's if you believe in the calculations of the seventeenth-century cleric James Ussher, the Archbishop of Armagh in Northern Ireland. Biblical scholars – among them, the Venerable Bede, and Dr John Lightfoot, Vice-Chancellor of the University of Cambridge – took delight in adding up the ages of the patriarchs of the Old Testament to ascertain the moment of Creation.

They all arrived at a remarkably consistent figure – but science was to prove them wrong. By the end of the eighteenth century, geological records on Earth were hinting that our Universe was a far more ancient place...

We are driving towards hallowed ground; up into the foothills above Pasadena, California. As we break through the LA smog, the dome hoves into view: home to the observatory that was to nail down the real origin of the Universe. Atop Mount Wilson is the 100-inch telescope (no-one calls it a '2.5 metre', out of respect for the venerated beast). We enter the observatory, and are stunned by its scale – today, astronomers build small domes in the interests of economy. It arches over an instrument built in 1917, which owes its design to the grand traditions of Victorian engineering. Think of a nineteenth-century bridge, constructed of mighty cross-trusses, pointing heavenwards. This awesome scaffolding is essential to support what was then the biggest telescope mirror in the world.

OPPOSITE *Pioneering solar astronomer George Ellery Hale, seen here focusing the Sun's spectrum at his Mt Wilson Observatory in Pasadena, California. Hale was instrumental in funding and building the world's leading two telescopes of their day – the 100-inch (2.5-m) at Mt Wilson, and the 200-inch (5-m) on Palomar Mountain, San Diego County.*

RIGHT *The Venerable Bede – one of many savants who attempted to estimate the date of the creation of the Universe. In this painting by the Victorian artist James Doyle Penrose, he is seen dictating his translation of St John's gospel into Anglo-Saxon.*

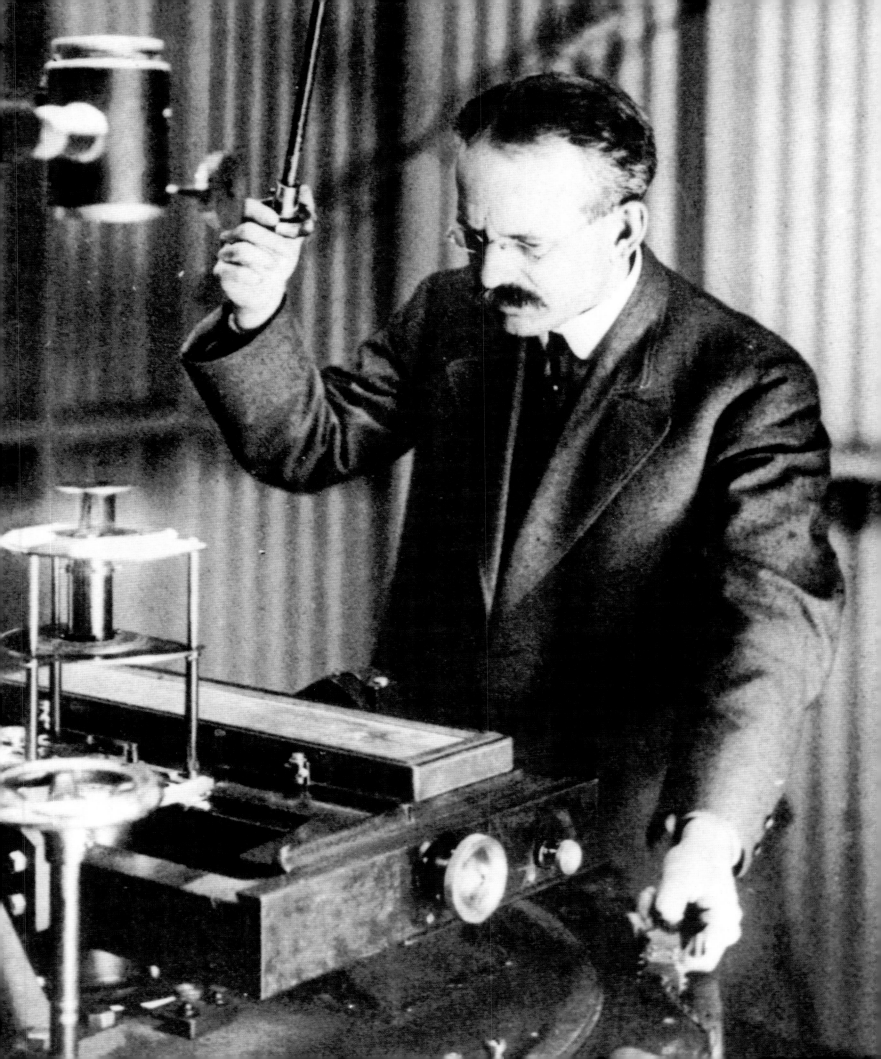

ABOVE *Hale's giant telescopes ushered in a new era in astronomy – the examination of the architecture of the Universe. As well as stars and planets, astronomers became aware that the heavens were filled with dark, natal gas – the stuff of future generations of stars.*

The 100-inch was one of the earliest 'light buckets' in the world to be used by professional astronomers. Until the end of the Victorian era, most professionals observed through lens telescopes – refractors – but there's a limit to how big a lens can be before it sags under its weight, and distorts the image. Replace the lens with a well-supported mirror, and the sky is literally the limit.

The brainchild of astronomer George Ellery Hale, the 100-inch is the instrument that ushered in cosmology: the study of the structure of the Universe – but not without some teething troubles. After having cast a 60-inch mirror for another telescope on the Mt Wilson site, Hale went for broke, and ordered the 100-inch to be cast at a factory in France. The first mirror had bubbles in it – and Hale suffered a nervous breakdown as a result. The next mirror broke when it was cooling. Hence the second nervous breakdown.

Eventually, the mirror was finished, figured, polished and incorporated into the telescope mounting. Hale's official biography, written by Helen Wright, describes the vicissitudes of the night of 'first light'. Hale, and his colleague Walter Adams, climbed the long flight of narrow black iron steps to the observing platform. The slit of the dome opened to a star-filled sky, and the telescope turned until it pointed to the brilliant planet Jupiter.

'As soon as the telescope was set on Jupiter, Hale crouched down to look through the eyepiece – desperately eager to know if all the years of effort had been successful. He looked, and said nothing. Only the expression on his face told of the horror he felt. Adams followed. His expression was a mirror of Hale's. They were appalled by what they had seen. Instead of a single image, six or seven overlapping images filled the eyepiece.'

In a situation like this, there's just one thing to do: go to bed and see if things improve when the mirror cools down to the ambient temperature, avoiding heat currents which can distort the view on the cosmos. But Hale couldn't sleep. Even trying to read a detective story didn't work.

'At 2.30 am, he returned to the 100-inch dome. Before long, Adams arrived – and confessed that he, too, had found sleep impossible. By this time, Jupiter was out of reach in the west. So they swung the great telescope over to the brilliant star Vega. Almost afraid to look, Hale again crouched down, and looked into the eyepiece. He let out a yell. The yell told Adams all he wanted to know. The telescope was an unqualified success.'

The 100-inch is a direct descendant of telescopes that were first pioneered over the other side of the Atlantic. These colossi, built by amateur astronomers in the eighteenth and nineteenth centuries in Britain and Ireland, opened up the Universe – taking humankind out of the realm of the planets, and plunging them into the uncharted territory of the stars. The trailblazer was William Herschel.

Although Herschel is chiefly remembered for his discovery of the planet Uranus, his obsession was actually with the wider cosmos. After he had doubled the size of the Solar System with his finding of the distant gas-giant in 1781, the science-obsessed King George III set up the great astronomer near the palace at Windsor. There, members of the royal family – and their guests from abroad – could enjoy scanning the heavens through Herschel's increasingly large telescopes.

His favourite was the '20-foot,' which had a mirror 19 inches (480mm) across. But he surpassed himself with the '40-foot.' This monstrous device boasted a mirror nearly 4 feet (1.2m) in diameter – but it was a nightmare to use. The observer had to be positioned high above at the eyepiece, while the telescope would be moved around by engineers who raised and lowered it with ropes and winches. It's alleged that a contemporary of Herschel's compared observing with the 40-foot to 'shaving with a guillotine' – and that some of the King's workmen went on strike, refusing to use it.

Herschel had hoped to use the 40-foot for his ultimate project: to make a really deep survey of the heavens. But in the end, he had to give up on his gargantuan construction: it suffered from tunnel vision. So he turned to his 20-foot to scan 700 selected areas of the sky.

He logged the stars in each region in a very methodical way, assuming that dimmer stars were further away than more luminous ones (this was before astronomers had managed to measure the distances to the stars). Herschel discovered that they were arranged in a grindstone-like pattern. His map bears a remarkable resemblance to images of the Milky Way Galaxy – our local city of stars – which telescopes reveal today.

As well as sweeping for stars, Herschel also noted the positions of nebulae: fuzzy patches in the sky. But these worried him. Towards the end of his life, he wrote: 'I must

BELOW *From 1917 to 1948, when Hale's 200-inch telescope saw first light, the 100-inch Hooker telescope – named after John D. Hooker, a wealthy Los Angeles businessman, and friend of Hale, who provided funds for its giant mirror – was the biggest in the world. In 1919, it was the first telescope to succeed in measuring the diameters of nearby stars.*

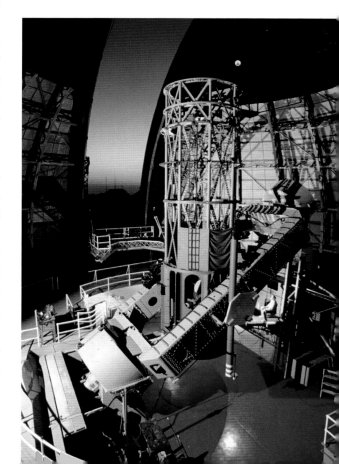

OPPOSITE LEFT *William Herschel's 1785 'grindstone' map of the Milky Way was a remarkably accurate portrayal of the appearance of our Galaxy. Even though astronomers had not succeeded in measuring the distance to a star by the late eighteenth century, Herschel expressed the scale of his model in 'siriometers' – his estimated distance to the brilliant star Sirius.*

OPPOSITE RIGHT *Compare Herschel's map of the Milky Way with this image of the nearby Andromeda Galaxy, which presents itself to us almost edge-on. They are both large spiral galaxies, though Andromeda is more than twice as big – the latest estimates put its diameter at 220,000 light years. Here, it is imaged with two of its brightest companions, M 32 (top), and NGC 205 (below).*

confess… my opinion of the arrangements of the stars has undergone a gradual change. We surmised nebulae to be no other than clusters of stars disguised by their very great distance, but a longer experience and a better acquaintance with the nature of nebulae will not allow a general admission of such a principle.'

It is sad that Herschel actually renounced his grindstone model in his later years: so close was he to discovering the structure of our Galaxy. His influence was such that other astronomers lost faith in it too. The eminent astronomer J.E. Gore observed: 'Sir William Herschel's disc theory, as it is termed, was abandoned by its illustrious author in his later writings and is now considered to be wholly untenable by nearly all astronomers who have studied the subject.' But by the middle of the nineteenth century, opinion was turning – largely due to the enthusiasm of the Third Earl of Rosse. Lord Rosse lived in Parsonstown (now Birr) in Ireland, and was a passionate amateur astronomer. Like Herschel, he delighted in building bigger and bigger telescopes – culminating in what was to become then the world's largest, with a mirror 72 inches (1.8 m) across.

Not everyone approved. Sir Robert Ball, Professor of Astronomy at Dublin and a great populariser of the subject, commented: 'I think that those who knew Lord Rosse well would agree it was more the mechanical processes incidental to the making of the telescope that engaged his interest, than the actual observations with the telescope when it was completed.'

Such assessments are unfair on Lord Rosse, who went on to make some extremely important observations of the mysterious nebulae with his 'Leviathan of Parsonstown'.

RIGHT *The 'Leviathan of Parsonstown' – Lord Rosse's mighty colossus, situated in Birr, central Ireland – was the largest telescope in the world from 1845 to 1917. Rosse's wife, Mary, was a talented photographer, who corresponded with the pioneering Henry Fox Talbot. She was also an expert blacksmith, designing the ironwork on her husband's telescope, and constructing the gates on their estate at Birr Castle – which stand to the present day.*

RIGHT *Lord Rosse's exquisite drawing of the 'spiral nebula' M 51. But what was the nature of the spiral nebulae? They were clearly different from amorphous clouds of gas, like the Orion Nebula. Were they part of our Galaxy – or something way beyond? Turn to page 233 to see a Hubble Space Telescope image of this beautiful object.*

Rosse used his huge telescope to home in on the nebulae. He drew beautiful pictures of them – and became fascinated by the nebulae that were spiral in shape.

What were the spiral nebulae? Were they part of our Milky Way? Or perhaps distant cousins of the Galaxy, millions of light years away in space? The development of photography at the end of the nineteenth century produced more detailed views of these denizens of the cosmos than ever before. When these photographs were first shown at the Royal Astronomical Society in the early 1890s, one Fellow recalled: 'One heard ejaculations of Saturn! – the nebular hypothesis made visible! – and so on.'

The debate now moves on to 1908 – to the Harvard College Observatory, where a young research assistant – Henrietta Leavitt – was poring over fragile glass photographic plates taken by a telescope in Peru. She was searching for stars that change in brightness. One colleague wrote: 'what a variable-star fiend Miss Leavitt is. One can't keep up with the roll of the new discoveries.'

These particular variables were in a star-system called the Small Magellanic Cloud. After identifying hundreds of variable stars, Leavitt discovered that some of them followed a striking pattern – the brighter the star, the longer it took to change in brightness.

The brightening and fading of these stars matched a naked eye star that had long been known to vary: Delta Cephei, chief of the clan of Cepheid variable stars. Leavitt's work showed that Cepheids could be used to find distances in space. In principle, the method was simple. Identify a distant Cepheid; and compare its brightness with a nearby Cepheid that varies in an equal period of time. Leavitt had proved that two

LEFT *The unassuming Henrietta Leavitt, whose discovery of the relationship between the brightness of a Cepheid variable star and its period of variation led to an understanding of the distance scale of the Universe.*

Cepheids with the same period must be equally luminous: so the faintness of the more remote star reveals its distance.

There was just one problem to begin with: astronomers didn't know the distances to any nearby Cepheids. But as they began to estimate the brilliance of these incredibly luminous celestial beacons, astronomers realised they could now get a handle on the most distant objects in the Universe.

One astronomer at the newly founded Mount Wilson Observatory, Harlow Shapley, declared of Henrietta Leavitt: 'Her discovery of the relation of period to brightness is destined to be one of the significant results of stellar astronomy.'

How right he was. Shapley had come to Mount Wilson as a result of a personal invitation from George Ellery Hale, and decided to study what exactly made Cepheids tick. But astronomy had not been his first choice. His real fascination in life was journalism – but on entering the University of Missouri, he discovered that the School of Journalism hadn't yet been built. He returned a year later – only to find that nothing had changed.

But he decided to hang around, and opt for a different course. He started going through the options alphabetically, and recalled: 'The very first course offered was A.R.C.H.A.E.O.LO.G.Y. ... and I couldn't pronounce it (although I did know roughly what it was about). I turned over a page and saw A.S.T.R.O.N.O.M.Y – I could pronounce it, and here I am!'

The newly-founded Mount Wilson Observatory attracted wacky characters like Shapley. Not least was Milton Humason, who dropped out of higher education in

BELOW *Harlow Shapley, then at the Mt Wilson Observatory, took Leavitt's observations a step further by measuring the distances to globular clusters – dense balls of stars that form a halo around our Galaxy. He discovered that the Milky Way was far larger then previously thought.*

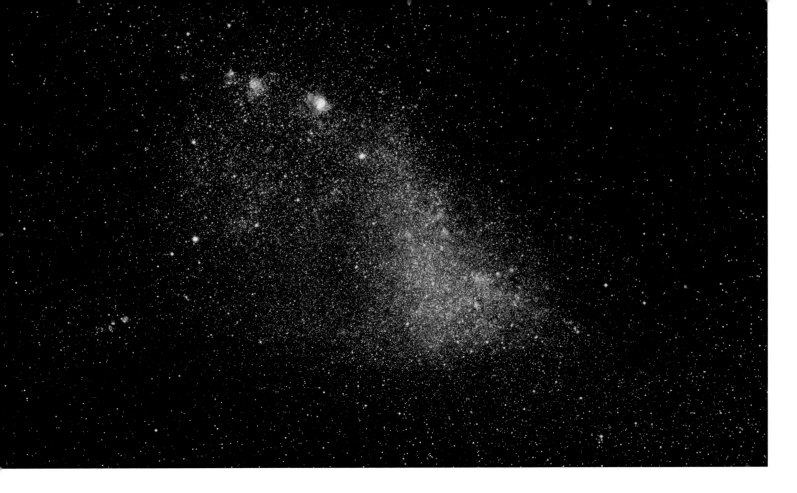

around 1910, and became a mule driver conveying wood and building materials up to the new telescope site. So keen was he on astronomy – although untrained – that he was next appointed janitor at the observatory. But his observing skills were superlative. He rapidly became a night assistant – and then a top astronomer in his own right.

And then there was Edwin Hubble. Like William Herschel, he had two careers: he was a brilliant lawyer, and a Rhodes Scholar at Oxford. (He was also, incidentally, an amateur boxer and a tank driver.) Hubble had dabbled in astronomy when he was at college, and eventually, the heavens took over from his legal career. He once said: 'I would rather be a second-rate astronomer then a first-rate lawyer. All I want is astronomy.'

Hubble was quickly snapped up by Mount Wilson to study the mystery of the nebulae. Starting on the 60-inch telescope, he found that some were definitely made of stars, rather than gas. Over the next few years, he busied himself with the new 100-inch. This powerful eye on the sky could pick out individual stars in the nebulae. Homing in on the nebula NGC 6822, he discovered 11 Cepheid variable stars. Using a combination of Leavitt and Shapley's work on these stellar beacons, he was able to measure a distance to the nebula. According to his calculations, it lay 700,000 light years away – far beyond our Milky Way Galaxy, which measures 100,000 light years across.

Spurred on by his findings, Hubble looked at other nebulae – particularly the spiral ones. He focused on the biggest, reckoning that they would be the closest. In M33, he found 25 Cepheids, and estimated it to be 850,000 light years away. M31 – the great nebula in Andromeda – yielded a very similar distance.

Today, we know from more accurate measurements, that these two objects lie nearly three million light years from us. But the point is that – almost overnight – there

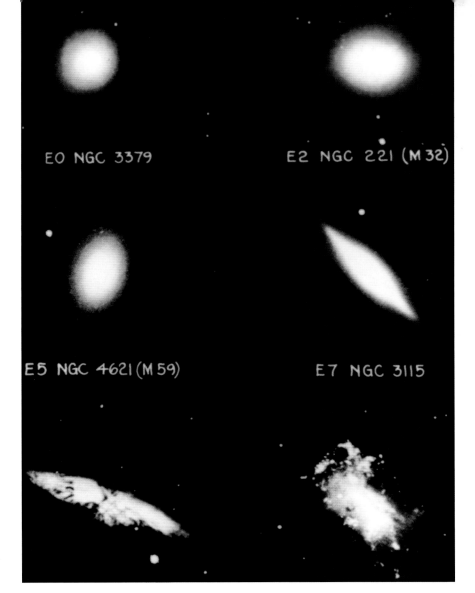

EO NGC 3379

E2 NGC 221 (M 32)

E5 NGC 4621 (M 59)

E7 NGC 3115

OPPOSITE *200,000 light years away, the Small Magellanic Cloud is a satellite of our Milky Way. Easily visible in the skies of the southern hemisphere, it is an active young galaxy, studded with pink regions of starbirth. Leavitt chose this galaxy to study the behaviour of Cepheids.*

LEFT *Hubble systematised his newly-found galaxies according to their shape and form. This page from his 'Hubble Classification' shows images of elliptical and irregular galaxies.*

BELOW *Hubble also discovered the fact that the Universe is expanding. These spectra of galaxies (in photographic negatives) show the 'H' and 'K' lines from calcium in their stars, which reveal increasing redshifts – the further the galaxy, the greater its light moves to the right (red) end of the spectrum.*

was a dramatic sea-change in the way we saw the cosmos. Far from being a local backyard of planets, or a parochial city of stars, our Universe had majestically extended its boundaries. At last, astronomers realised that many of the 'nebulae' were independent star-systems outside our Galaxy. Hubble called them 'extragalactic nebulae' – but today, we simply call them galaxies.

Like many scientists of the Victorian and post-Victorian era, Hubble was into classification in a big way. He photographed his galaxies – and discovered that not all of them were spiral. Some were indeed beautiful Catherine Wheels of coiled-up stars, as we know is true in the case of our Milky Way. These galaxies are brimful of young stars and pregnant nebulae tracing out the spiral arms.

Then there were elliptical galaxies: baleful balls of old red stars, with no natal gas, and well past the galactic menopause. Dwarf galaxies completed the set – ragged little irregulars, frantic with starbirth, and dwarf ellipticals, which had probably collided with their host galaxies and been stripped of their building materials in the process.

Now it was time to analyse the light from Hubble's newly-found galaxies by capturing their spectra. Some of his work had already been done for him: Vesto Melvin Slipher at the Lowell Observatory in Flagstaff, Arizona (where Pluto would be discovered in 1930) looked at the spectra of several 'spiral nebulae' in the early years of the twentieth century.

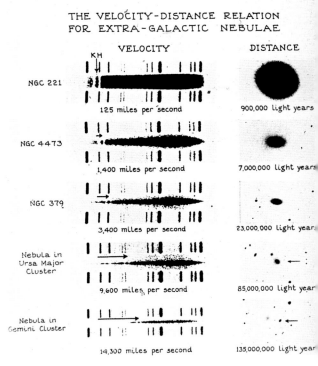

THE VELOCITY-DISTANCE RELATION FOR EXTRA-GALACTIC NEBULAE

VELOCITY   DISTANCE

NGC 221
125 miles per second   900,000 light years

NGC 4473
1,400 miles per second   7,000,000 light years

NGC 379
3,400 miles per second   23,000,000 light years

Nebula in Ursa Major Cluster
9,600 miles per second   85,000,000 light years

Nebula in Gemini Cluster
14,300 miles per second   135,000,000 light years

ABOVE *The Hubble Deep Field: these images of distant fledgling galaxies in Ursa Major were captured by the Space Telescope, named after the great astronomer.*

BELOW *Georges-Henri Lemaître (centre) was the first to suggest, on theoretical grounds, that the universe is expanding. Here he relaxes with colleagues Albert Einstein (right) and Robert Millikan (left) – famed for his 'oil-drop' experiment, which measured the charge on the electron.*

Most striking was the fact that the 'nebulae' appeared to be running away from us. Picture a scene in a busy city street: a police car sweeps past with its siren wailing. On approach, the siren sounds higher-pitched; after it has passed, the pitch drops. That's because the wave-fronts from the siren are bunched up towards us as the car approaches – but they trail behind once the car has sped on its way.

And so it is with light. A mixture of colours from short-wavelength blue to long-wavelength red, the spectra of approaching objects shifts towards the blue; while those of receding objects shift towards the red.

In 1928, Hubble – assisted in no mean way by Milton Humason – got to grips with measuring the 'redshifts's of their galaxies. Humason did the donkey work (no pun!), and discovered that virtually every galaxy was receding from our own. The further the galaxy, the greater was the velocity of the recession. There were many scratched heads at that point. Was light getting 'tired' and losing energy by travelling the colossal distances of space? Or was something more fundamental going on?

The answer came in 1931 from Father Georges-Henri Lemaître – a Belgian Roman Catholic priest, and astronomer. The Universe is expanding, he declared, in a paper published in the journal *Nature* on the 'hypothesis of the primaeval atom.'

Essentially, Lemaître had hit on the origin of the Universe. It began as a tiny entity, then expanded – creating the stars, planets and galaxies that we observe today. Theoreticians like Willem de Sitter had been working with Albert Einstein's equations of Relativity, and had also come to the tentative conclusion that our Universe was expanding. But Einstein – while he approved of the mathematical arguments of Lemaître's idea – couldn't abandon the idea of a static cosmos.

Following a series of international conferences in the early thirties, however, Einstein and Lemaître were able to exchange views frequently. Einstein was beginning to become won over. Finally, in 1935 – at Princeton in California, where Lemaître gave a presentation – Einstein stood up, applauded, and is alleged to have said: 'This is the most beautiful and satisfactory explanation of creation to which I have ever listened'.

Billions of years ago – out of nothing – a tiny speck of brilliant light appeared. Almost infinitely hot and dense, inside this fireball was the whole of space. And with the creation of space came the birth of time. The great cosmic clock started to tick.

The energy in this primeval fireball was so concentrated that matter spontaneously started to appear. Einstein himself had predicted that this was possible: his famous equation $E=mc^2$ says that mass (m) and energy (E) are interchangeable ('c' in the equation is the speed of light). The first matter would have taken the form of subatomic particles like electrons, positrons, quarks, WIMPs, cosmic strings and primordial black holes, which cannoned around like microscopic billiard balls. These were the building blocks of the stars, planets and galaxies we see around us today.

But were they? Ever the maverick, Fred Hoyle objected to the new orthodoxy. Aided and abetted by his colleagues Hermann Bondi and Tommy Gold, Hoyle – in 1948 – came up with a completely different theory for the origin of the Universe: the Steady State. In their view, the Universe had no end, and no beginning. Although expanding, the cosmos stayed in perfect balance, like a washing-up bowl that's kept topped-up by a trickle from a tap. The 'tap', in this case, was the continuous creation of matter from energy – albeit at the paltry rate of about one new hydrogen atom in a volume the size of a wine bottle every billion years.

Hoyle was always disparaging of Lemaître's theory. He nicknamed it 'the Big Bang', which – ironically – only served to popularise it. But Geoff Burbidge – who with his wife Margaret, worked with Hoyle on the origin of the elements in stars – still believes in a modified Steady State theory. 'The classical Steady State doesn't work', he admits. 'It's got too much observational evidence against it. But Fred and I in the 1990s worked together, and – what you've got to take very seriously – there's a lot of evidence that the centres of galaxies are where there are large-scale explosions. Matter and energy are pouring out of these things'.

Back in 1948, a rival camp was setting up – in favour of the Big Bang. The colourful Ukrainian physicist George Gamow (who defected successfully to the United States with his physicist wife Lyubov – after first attempting to escape by kayaking across the Black Sea to Turkey), teamed up with two colleagues to promote their ideas.

The team comprised Gamow, Ralph Alpher (Gamow's student), and Hans Bethe from Cornell University, New York. Actually, Bethe had nothing to do with the research: Gamow, with his ironic sense of humour, decided that it would be fun to have a paper authored by Alpher, Bethe and Gamow – after the first three letters of the Greek alphabet: alpha, beta and gamma.

ABOVE *A stunning Hubble Space Telescope image of the spiral galaxy M51 – 'The Whirlpool'. Its curving arms are laced with the delicate traceries of starbirth. Companion galaxy, NGC 5195, lies slightly behind the cartwheel-shaped M51. The smaller galaxy blundered into The Whirlpool some 500 million years ago, triggering a burst of star formation.*

As well as suggesting how the earliest elements – hydrogen and helium – formed in the Big Bang, Gamow made a prediction: that there should be an 'afterglow of creation.' Space, he maintained, should have a temperature of around five degrees above absolute zero, as a result of the residual heat from the cosmic inferno. It would be 16 years before his forecast proved to be correct.

Meanwhile, the young discipline of radio astronomy – a direct descendant of radar operations during the Second World War – was growing up. This was to prove to be decisive in the debate about the origins of our Universe.

In Britain, two establishments rose up in the rural fields: one in Cheshire, the other just southwest of Cambridge. Giant radio dishes, looking out on the distant cosmos, started to become a feature of the English landscape.

In charge of the Cambridge facility was Martin Ryle – a member of the distinguished Ryle family, and nephew of the renowned Oxford philosopher Gilbert Ryle. After taking his physics degree at Oxford, and working on radar during the War, Ryle and his team founded the Mullard Radio Astronomy Observatory – part of the Cavendish Laboratory at the University of Cambridge.

Ryle was a fascinating character. He was brilliant – probably one of the most influential astronomers of the twentieth century – but secretive. He didn't want his valuable results to emerge in public. He was also difficult to work with, as one of us (Nigel) found when researching in his group! And his temper was legendary.

Having said that, we personally find it inspiring that scientists like Ryle and Hoyle were around in our young years to make the impact that they did. After all, what is it

ABOVE *Mavericks of the cosmos: the young Hermann Bondi (centre), Fred Hoyle (right), and Tommy Gold (left) get up to mischief at an astronomy conference. The trio, who set forth the Steady State Theory as an explanation for the origin of the Universe, always defied the conventional. Note the lack of their name badges!*

OPPOSITE BELOW *Champion of the Big Bang: talented writer and physicist George Gamow. His 1948 paper explained how only a violent origin could produce the amounts of hydrogen and helium that we observe today.*

ABOVE *Centre of the Virgo cluster of galaxies. In the 1950s, the burning question was: had galaxies always existed? Or were they born in fire in the Big Bang?*

that drives science? Is it the question of issues, confronted by enormous teams of international researchers? Or can individuals drive the process ahead? We like to think that the latter is possible.

Our former Professor at the University of Leicester – Jack Meadows – sets the scene. He moved from conventional astrophysics into the fields of history of astronomy and information technology. 'I started out my career in the United States, about 1960,' he tells us. 'There were still senior astronomers who'd been brought up on the old tradition. They worked by themselves, did their own observations, and then they published their work.'

'They were a little unhappy about the growth of teamwork, because they thought that perhaps individual flair wouldn't be allowed to exert itself to the same extent. And they also thought that people in teams tended to forget what had gone before. One senior astronomer – Otto Struve – said that he didn't think that most graduate students knew any astronomy more than ten years back.'

Teamwork, of course, is dictated by consensus. Geoff Burbidge makes the point: 'There has always been consensus. For money to do science, you need grants, and you need observing time for astronomy. Everything is reviewed, and the reviewers are all of the majority opinion. So – as a result – no observation has ever been made with the Hubble Space Telescope to test out any ideas that are unorthodox.'

However, back in the 1950s, the old traditions of astronomy were still hanging on – teams were, of necessity, getting together to build the eyes on the sky for the new millennium. But personalities were still crucial. And Ryle and Hoyle exemplified this to an unsurpassed degree.

The contretemps between the two scientists was spectacular, to say the least. It was conducted in public, with blunt-talking Yorkshireman Hoyle confronting the elegant and aristocratic Ryle.

Ryle's radio telescopes could actually see further into the night sky than any conventional optical telescope of the day. What he discovered was that – the further into the cosmos he looked – the more closely were galaxies gathered together.

Because of the time it takes light to cross the cosmos, looking into the distance means looking back in time. There was just one inescapable conclusion: that the Universe was more compact in the past, and has expanded to its present size.

It was an almost fatal blow for the Steady State.

After the dust between the two scientists had settled down, media-wise, the mischievous George Gamow waded in. Not only the prolific author of hundreds of papers and textbooks, he also wrote the wonderful series *Mr Tompkins*, in which his (self-drawn) cartoon character explores the world of science.

But Barbara, his second wife, was as equally talented as her husband in the literary field. After the debacle in Cambridge, she couldn't resist penning this ditty – meant to be sung to the tune of *O Tannenbaum*.

But the Steady State theory didn't seriously hit the buffers until 1965. And at this point, the scene shifts from Cambridge to Holmdel, New Jersey – to the Bell Telephone

Laboratories site. Now: enter two very confused astronomers – Arno Penzias and Robert Wilson.

ABOVE *Dedicated astronomer Martin Ryle in 1957, working on recording equipment that monitored radio waves. His research convinced him that more distant galaxies were packed more closely together – evidence that the Universe was created from small beginnings, in a Big Bang.*

ABOVE *Discoverers of the afterglow of the Big Bang: Arno Penzias (left) and Bob Wilson (right) stand in front of the horn antenna in Holmdel, New Jersey, where they first accidently tuned in to the cosmic microwave background in 1965.*

'We frankly did not know what to do with our result,' despairs Penzias, 'knowing that no astronomical explanation was possible.' Hardly a promising start for a finding that ranks right up there with the discovery of the expanding Universe. It went on to earn the team a Nobel Prize for Physics in 1978 – and world fame as the duo that finally nailed down the reality of the Big Bang.

Any thoughts of fame or discovery were far from the minds of the team when they started their research project in the early 1960s. Wilson recalls: 'We weren't setting out to measure the properties of the Universe, but the properties of our own Milky Way'. The two astronomers were hoping to detect radio waves coming from the tenuous spherical 'halo' that surrounds our disc-shaped Galaxy – a measurement that required an especially sensitive radio telescope.

Penzias and Wilson settled on an unusually-shaped antenna, like a giant metal horn 20 feet (6 m) across at the wide end. It had been built to receive messages from the first primitive communications satellites.

'When we first turned it on, we knew immediately there was a signal in there that we didn't understand', remembers Wilson. The astronomers looked to all possible terrestrial sources. 'Maybe it was junk from New York City.' says Wilson of their first suspect – radio interference from one of the most electronically-bristling cities in the world.

However another explanation lay closer to home – as Robert Wilson vividly recalls. 'There were a pair of pigeons living in the antenna, so the inside of the horn was covered in pigeon droppings. Arno and I in our white lab coats got up there with a broom and cleared out all the droppings, but nothing seemed to change things.'

After a year, Penzias and Wilson were forced to conclude that the origin of their rogue signal lay in the sky. And if that was the case, it was spread out with a fantastic degree of uniformity. It showed no change with day, night or time of year, and corresponded to a constant temperature of about three degrees above the absolute zero of the temperature scale. It was as if the whole Universe was very, very slightly warm.

Arno Penzias mentioned their baffling findings to a physicist friend, Bernie Burke. It rang bells. Burke knew of work going on at nearby Princeton University, led by a theoretical physicist called Robert Dicke, on the origin of the Universe.

Wilson was stunned. 'When we talked with Dicke, we found out that they were looking at a theory of the Big Bang in which our Universe began. In such a situation, the Universe would be hot – but if it was hot in the beginning, it would cool down, and the radiation that filled it at the beginning would be visible now as radio waves'. Dicke and his colleagues had also calculated how hot the Universe should be at this present time – and their answer was a few degrees above absolute zero: exactly as George Gamow had predicted so many years before.

The Princeton team came out to Holmdel. They had actually been on the verge of constructing a radio antenna themselves to check out their theoretical prediction. 'But when they looked at our experimental apparatus', says Wilson, 'they almost immediately agreed that we'd made the measurement they were hoping to make.'

What Penzias and Wilson had stumbled across in 1965 was the afterglow of creation itself – the cooled-down relic of the inferno in which our Universe began. Coupled with Edwin Hubble's earlier discovery of the expanding Universe, this finding left astronomers in no doubt about the nature of creation: our cosmos was born in a Big Bang.

After the duo had made their discovery of the legacy of the Big Bang, other groups of astronomers followed up with even more detailed observations – looking for structure in the 'Microwave Background'. They realised that they were looking at – essentially – a template for the building of the future Universe. The smoothed-out radiation from the Big Bang, spread so uniformly over the sky, ought to contain knots and wrinkles that would point astronomers to concentrations of matter destined to give birth to baby galaxies. But they could see none – at least, not from the surface of planet Earth. The answer was to travel into the clear, unwavering skies of space.

Cosmologist Carlos Frenk of Durham University takes up the story: 'In 1992, the COBE satellite took a picture of the baby Universe, and what it saw were precisely those

LEFT *Confirmation of the Big Bang: the 4C Array at the Mullard Radio Astronomy Observatory, 5 miles (8 km) south-west of Cambridge, conducted a survey of almost 5000 galaxies in the early 1960s. It revealed the architecture of the Universe, proving that our evolving cosmos must have come from a violent past.*

RIGHT *This image from the WMAP satellite (Wilkinson Microwave Anisotropy Probe) peers into the dark origins of our Universe, only 300,000 years after the Big Bang. This view shows the whole sky mapped onto a 3D sphere, like the star-globes of medieval times. The colours in this image show that structure is already starting to form in our young cosmos: red and yellow regions will become fledgling galaxies – concentrated around cores of dark matter.*

small imperfections, which we now understand are nothing other than the precursors of the great galaxies and great galaxy clusters that we see in the Universe today. These small irregularities are the missing link that connects the very early cosmos and the Big Bang with our mature Universe of galaxies.'

The COBE satellite was the ultimate thermometer. Designed to probe the chill temperatures of deep space, its goal was to seek out the irregularities that would presage the formation of galaxies. Seeds of galaxy formation would reveal themselves as cooler, denser knots in the dense fog – so-called 'ripples in space.'

COBE came up trumps – and its findings made front-page news. 'Had it not been for the small imperfections, the Universe would be completely boring – there'd be nothing in it,' observes Frenk. 'But these irregularities, these imperfections, grew into galaxies, into stars, into planets, and eventually into people. We can say that we are nothing but the descendants of these very small imperfections that were imprinted in the Universe a few instants after the moment of the Big Bang.'

The satellite – and its successor, WMAP – have not only scrutinised the texture of our fledgling Universe: they have pinned down the date of the Big Bang itself. For decades, debate raged around the issue. But the detailed measurements of these spacefaring explorers have at last told us the date of creation: it was 13.7 billion years in the past.

And they have also told us more about the construct of the Universe. When we think about the heavens, we envisage a dazzling cosmos filled with brilliant stars, glorious galaxies, and softly-glowing nebulae. But nothing could be so off-the-mark.

Over 90 per cent of our Universe is invisible – filled with particles of mysterious dark matter. And astronomers have no idea what it is. Theoretical physicists working on the kinds of particles produced in the Big Bang say that dark matter cannot be anything ordinary – it has to be something very exotic.

Carlos Frenk from Durham observes: 'In my opinion – although we still don't know what the dark matter is made of, we do know that it controls the Universe.' Its gravitational attraction determines what structures won't form, and which structures will.

Dark matter also dictates the future of the Universe: in one of life's ironies, our fate is actually determined by material we cannot see. Will the cosmos continue to expand, or eventually collapse under the pressure of its own gravity? 'Dark matter really is the master of the Universe.' concludes Frenk.

But, as we look towards the future of the Universe, there has been a recent, unexpected sting in the tail. By observing distant supernovae – exploding stars which act as standard cosmic beacons in the recesses of the cosmos – astronomers have now discovered that the Universe is not just expanding: it is accelerating.

What could be causing the Universe to accelerate? Says Carlos Frenk: 'We thought that gravity was really the only player, but now we're beginning to suspect that there might be another character – a new, mysterious force – that could play a role in determining the long-range future of our Universe. Einstein knew all about it – he was the first person to consider there might be a repulsive force in the Universe to counteract the effects of gravity.'

That force is 'dark energy' – another unknown for astronomers to grapple with. And the discovery of the accelerating Universe is so new and unexpected that scientists are having to scramble to come to terms with it. But it at least means that we can put our bets on living in an increasingly dark, lonely and cold cosmos.

In the light of these new discoveries, Astronomer Royal Martin Rees looks forward to the very far future. 'Suppose we look ahead to when the Universe is a hundred billion years old. It will be a dull and dark place, because all but the faintest and most slow-burning stars will have died, leaving dead remnants like white dwarfs, neutron stars, or black holes. Now, let's go ahead to when the Universe is, say, a trillion, trillion years old: all the stars will have died, and the Universe would be very, very dispersed indeed.'

But even this would not be the final outcome, for atoms would still remain. 'Atoms don't live forever.' says Rees. 'They gradually decay. If you waited a number of years – actually, one followed by about 35 zeros – all the dead stars would erode away. You'd still have dark matter and black holes. But we believe that even black holes don't live forever, and eventually evaporate.'

What about the unimaginably far future? 'When the Universe is so old that – in years – it's one followed by 100 zeros, even the biggest black holes will probably have gone. Then the Universe will just be very, very dilute radiation and dark matter, plus nothing else but a few electrons and a few positrons, and that's all. And in that state, the Universe can go on expanding for the infinite future.'

What would Archbishop Ussher have made of all this, we wonder?

ABOVE *Martin Rees, England's Astronomer Royal, looks to the future of our expanding Universe. It is a bleak prospect: our cosmos will continue to expand, but all of its matter will finally decay.*

BELOW *Archbishop Ussher of Armagh – who decreed that the Universe was born on Sunday, October 23, 4004 BC – would have been amazed at our recent discoveries about the cosmos. These findings prove that science – and not speculation – has the final say.*

Violent Universe

February 1942. The world was at war. But little did anyone realise that the violence of combat would alert us to violence on an altogether greater scale – the recognition that our Universe, far from being a calm haven, is a seething cauldron of unimagined energy and destruction.

It all began with the development of radar: echoes of radio waves that now allow air traffic controllers to track airplanes. In WWII, the technique was in its infancy; but it was being used – with some success – to home in on enemy bombers and ships.

One of the radar experts was 33-year-old Stanley Hey – a brilliant young physicist who had signed up for war service on the government's scientific register. But there was always a worry that the enemy might 'jam' the Allies's radar by transmitting powerful radiation on the same wavelength.

Hey later recalled: 'Scientifically, the study of jamming appeared to be a most unattractive proposal – but it was wartime. Little did I realise that it would become an exciting and intriguing phase in our work.'

His first task was to collate reports of jamming or interference from all the British radar sites. 'I recall one morning, after preparing my report ready for typing, a dispatch rider appeared. I was wanted urgently at the War Office – a car was coming for me at once'

Hey had no time to change from his old sports coat and flannels, before having to testify to the Major, the Brigadier, and finally to the Director-General of the Royal Artillery. 'Although I must have seemed to be a rather shabbily dressed young man, my opinion was sought with great respect'.

On February 27 and 28, Hey discovered powerful radar jamming – during the hours of daylight – covering the whole country. 'It was fortunate that there was no

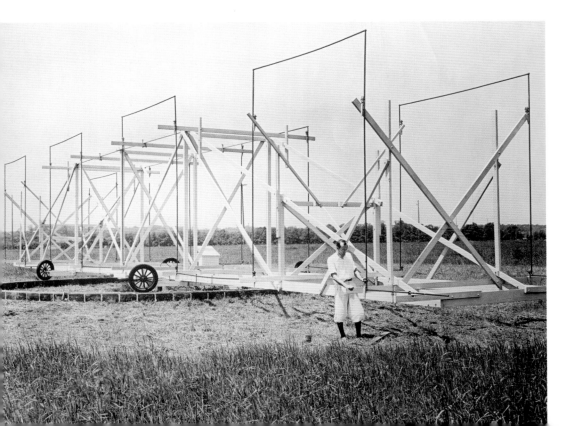

LEFT *1931: Karl Jansky stands in front of the great rotating antenna which he built to detect terrestrial static that might interfere with transatlantic communications. Instead, it discovered radio waves coming from the distant band of the Milky Way.*

air-raid in progress', he observed. 'Had there been, telephone lines would have been white-hot with panic calls.'

After analysing the data from several radar stations, Hey came to an inescapable conclusion. The jamming was not coming from a Nazi secret weapon, but from the Sun. He checked with the Royal Greenwich Observatory, and discovered that there was an enormous sunspot group generating powerful magnetic eruptions – solar flares – that spew streams of energetic electrically-charged particles into the Solar System.

It would be an intimation that we live in a violent Universe.

But the first radio waves from space had been discovered as long ago as August 1931. Twenty-six-year-old Karl Jansky was a physicist and engineer employed by the Bell Telephone Laboratories at Holmdel, New Jersey (where the afterglow of the Big Bang would be detected, later in the century).

Bell Labs were keen on setting up a transatlantic communications system, and wanted Jansky to investigate sources of interference. He built a huge radio antenna, which detected static from nearby thunderstorms, distant thunderstorms – and an unknown hiss that prevailed in the background. Delving deeper, Jansky discovered that it came from the Milky Way.

ABOVE *A huge coronal loop arches above two sunspot groups on the Sun's surface. The incandescent gas, at temperatures of around a million degrees, traces the powerful magnetic field linking the groups. At the edges of the loop, charged particles are spewed at colossal speeds into space, generating radio waves which – during World War II – were taken to be enemy signals jamming British radar.*

ABOVE *Bernard Lovell, founder of Jodrell Bank, photographed in 1959 at his desk in the then-named 'Experimental Station'. Behind him is his leviathan – for many years, the biggest radio telescope in the world.*

BELOW *A thoughtful Stanley Hey – then in his late seventies – reminiscing on his discovery of radio waves from our Sun. Hey pioneered radio astronomy in Britain, leading to the knowledge that we live in a violent Universe.*

It could have been the start of radio astronomy. But Bell Labs reckoned the static posed no threat to their communications system – and took Jansky off the job. He never returned to astronomy again.

Jansky's discovery had however fired up another young enthusiast, Grote Reber from Illinois, who built the world's first purpose-designed radio telescope in his back yard. Reber conducted a pioneering radio survey of the sky, but it was largely ignored because America was in the depths of the Great Depression – and Reber was an amateur.

However, time has a way of healing wounds. Reber left the States for Tasmania in the 1950s, when he realised that he had been overtaken by professional radio astronomers. He died in 2002 – and his ashes are scattered at 24 major radio observatories around the world.

Stanley Hey was well aware of Reber's work. 'His effort was quite extraordinary, because he constructed his own radio dish that was 31 feet (9 m) in diameter.' Reber inspired Hey and his colleagues to make their own radio map of the sky.

Using the aerials at his Army establishment at Richmond Park in Surrey, Hey and his team surveyed the heavens. 'A new and exciting discovery emerged', Hey recalled. 'It was radiation from a small region in the direction of the constellation of Cygnus'. Understated as ever, Hey was not to realise that he had found the first exploding galaxy – Cygnus A.

We have never met such a delightful individual as Stanley Hey. He was a true English gentleman, and so modest about his achievements. Very little is known about him – nor of his inspiring fellow-scientist wife, Edna – he even entitled his privately published autobiography *The Secret Man*.

After the War, Hey joined the Royal Radar Establishment in Malvern, England. They had no radio astronomy research in progress, and so he took it upon himself to rectify the situation. As Hey recalls, a giant German Wurtzberg radar dish had been 'acquired' by the Establishment. Hey used it as the basis to build the first proper radio telescope in Britain, with a collecting dish 45 feet (14 m) across. He quietly carried on his research out of the glare of publicity.

But his reputation had created waves amongst the young war scientists. Among them was Martin Ryle, a researcher at Cambridge who wanted to study radio waves from solar flares in more detail. And at Manchester, another budding radio astronomer, Bernard Lovell, borrowed Hey's equipment to study radar echoes from meteors – another of Stanley Hey's discoveries.

Lovell recalls searching for a site for the equipment that was free from electrical interference. 'When I tried it out on the university campus, there was masses of interference from electric trams. I asked the university if they had a few acres in a quiet spot and they said – yes: the botanists have a research ground about 20 miles (32 km) south of Manchester. They gave me permission to go there for two weeks. And that was in December 1945.'

Over 60 years later, the radio telescopes at the botanical gardens of Jodrell Bank are still going strong – embellished by a beautiful arboretum which surrounds the dishes.

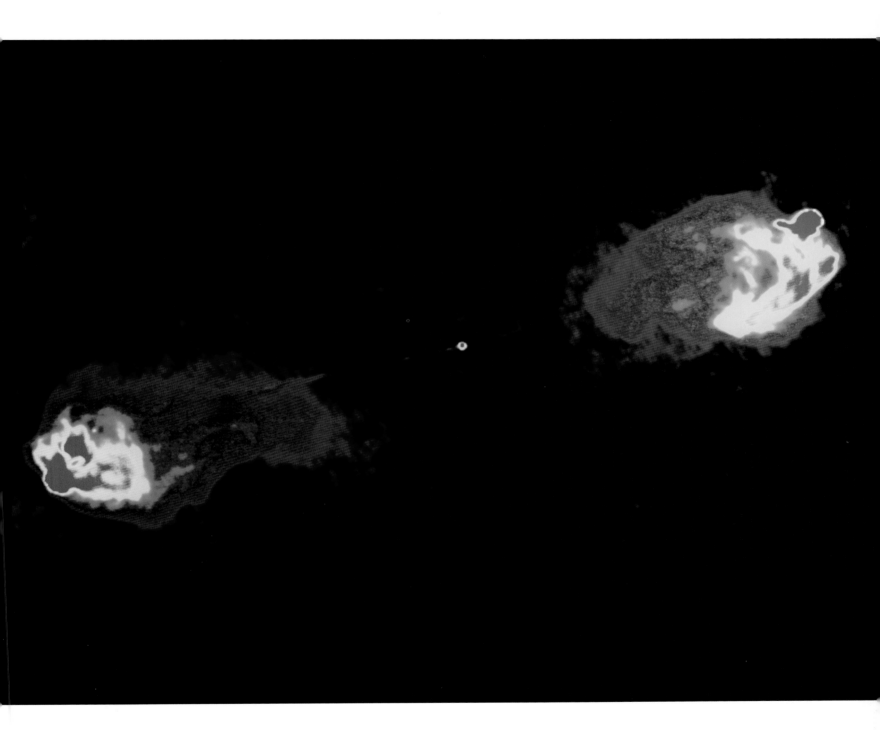

ABOVE Hey's group made the first discovery of radio waves coming from outside our Galaxy. A small source, which they called Cygnus A, coincided with what appeared to be two colliding galaxies. Now astronomers know that there is just one supermassive galaxy in that position, gashed by a thick dust lane. The radio waves come from two enormous clouds of high-energy electrons, which extend for half a million light years. These are produced by powerful jets spewed out of the galaxy's core.

Cygnus A is 700 million light years away, yet it is the second-brightest radio source in the sky. Its radio output is a million times that of our Milky Way.

Like many astronomers, Lovell has other passions. He has created not just one, but two arboretums. His other loves are classical music –'I was a church organist until I had to give up through bad eyesight a few years ago. And cricket. And, of course, trees.'

'I'm speaking from the arboretum that surrounds my home, which my wife and I created. It has quite a rare collection. I have at least one sample of almost every genus of wooded plant according to the reference books. In the British Isles, I'm only about three dozen short.'

'In my old age – well: my ideal day is to spend the morning working in the arboretum, the afternoon at Jodrell, and the evening to listening to a performance on Radio 3.'

Lovell's life was not as tranquil in the past. In the early 1950s, he embarked on an audacious project to build what was then the largest radio telescope in the world. At 250 feet (76 m) in diameter, it would be the astronomical equivalent of the Forth Railway Bridge – and a leviathan in the world of engineering.

In collaboration with the Sheffield engineer Charles Husband, Lovell designed an architectural landmark that rose from the Cheshire plains, and which would later become an international icon.

What led him to these gargantuan ideas? 'Well, I tried, you see', explains Lovell: 'I'd been working for the Royal Air Force and it seemed nothing to me to put an enormous

scanner under the belly of a Lancaster and fly it at 30,000 feet (9000 m). I thought to build this thing which I could steer to any part of the sky would be a simple matter.'

But it was not to be. With advances in radio astronomy going on all over the world, Lovell kept changing the design of his cherished project. 'We ran into a debt of £250,000 in '55, and that led to the investigation by the Public Accounts Committee. It was actually the *Sputnik*, in '57, that saved us.'

'We rapidly converted the telescope and put a transmitter on it – to obtain radar from the *Sputnik* and its launch vehicle. That's how we got signals from the carrier rocket, which was an intercontinental ballistic missile. And that saved us, because the Press – who had been very antagonistic – rallied around in support.'

Later, at the peak of the Cold War in 1960, Lovell had a private visit from an American General. He wanted Lovell's assistance to track what the General claimed to be rockets that the US were launching towards the Moon and Venus. The Americans didn't have the technology to track them – but Lovell did.

'After the first rocket was launched,' remembers Lovell, 'We sent out a signal that fused the bolts. The newspapers were full of this. And I had a call from Lord Nuffield asking 'How much is owing on that telescope of yours?' And I said, oh – about £80,000. He said: 'Oh well. We'll send you a cheque.'

'I tried to thank him, and he said: 'that's all right, my boy – you haven't done too badly.' And that was the end – that was the fairytale end to all these years of problems.' Lovell's powerful dish complemented what Martin Ryle was achieving, away from the scrutiny of the Press. In Cambridge, Ryle was scouring the sky to pick out as many radio sources as he could, and published them in three catalogues. By 1959, his group had discovered over 2000. Many seemed to be peculiar galaxies, like Cygnus A, now dubbed 'radio galaxies.'

But Lovell's team, headed up by Henry Palmer, were finding that some of these radio sources were unexpectedly small.

In California, a young Dutch-American astronomer, Maarten Schmidt, set himself the task of finding out what these unexplained denizens of the Universe actually were. Instead of using a radio telescope, he was observing what was then the largest optical telescope in the world – the Hale 200-inch (5-m) reflector on Palomar Mountain, California.

As kids who grew up reading about this telescope, it is nothing short of a pilgrimage to visit Palomar. The observatory is a cathedral to the cosmos. Its vast Art Deco dome is covered in hundreds of softly-glowing aluminium tiles. We look up at the mighty 200-inch (5-m) and gaze in awe at its incredible engineering – and, as at Jodrell Bank, the echoes of Victorian bridge-building are not far away.

At the top of the iron-latticed tube is a cylindrical observer's cage, where an astronomer would sit for perhaps eight hours at a time – either capturing images, or analysing the light from the most distant bodies in the Universe. And that's exactly what Maarten Schmidt was doing over the winter of 1962-63.

Schmidt took spectra of three of the mystery objects from the third catalogue of the Cambridge team, including 3C 48 and 3C 273. He imagined that these point-like

BELOW *Maarten Schmidt – discoverer of quasars – next to his cherished old-fashioned microscope with which he measured the redshift in the spectrum of 3C 273. Schmidt keeps this treasured tool proudly on view in his office at Caltech, Pasadena.*

objects were stars – but couldn't make head or tail of their spectra. 3C 273 also appeared to be double.

Schmidt scratched his head and – once back at Caltech – kept returning to his plates. In the case of 3C 273, a relatively bright object, there was a vaguely familiar pattern to the few spectral lines that he had been able to capture. He thought: 'I was sort of irritated. I said, darn it – I'll take my slide rule and I'll show that these lines are getting regularly spaced.'

Then it hit him. 'Jesus, a redshift. And immediately it struck me how incredible it was.'

Stars don't have significant redshifts. Only distant galaxies, whose lightwaves are lengthened by the expansion of the Universe, show the characteristic patterns of the Doppler Effect. And the redshift of 3C 273 was 16 per cent – indicating a distance of almost 2.5 billion light years.

ABOVE *The Hale 200-inch (5-m) telescope on Palomar Mountain, California, where Maarten Schmidt first captured the light from 3C 273. In the past, astronomers would spend all night in the cramped observer's cage at the top of the telescope. The observer here is Edwin Hubble – who discovered the nature of the galaxies and the expansion of the Universe.*

Schmidt was astonished. The object was 40 times brighter than a normal galaxy, but appeared no bigger than a star. 'Phew. That was something to think about. Anyway, in the ruckus I made some noises and my colleague Jesse Greenstein came by.'

'He said, my God, and produced his paper on 3C 48. Within 15 minutes, we found that its red shift was 37 per cent.'

Which meant that the object was nearly four billion light years away.

'Now we remembered that 3C 48 varied in time – so now it gets even wilder: a star that's actually a galaxy that's forty times brighter than the biggest ones known, which varies within months.'

The objects became known as 'quasars' – 'quasi-stellar radio sources.' The name was basically an admission that astronomers didn't have a clue as to what they were. All that they could surmise were that quasars were extremely small, dazzlingly bright, and very distant.

Meanwhile, a young research student with the radio group at Cambridge – Jocelyn Bell (now Bell Burnell) – was about to embark on her own study of quasars. Now a professor based at Oxford, Bell Burnell was not quite expecting what she was going to find…

BELOW *Since their discovery, the nature of quasars has been prised open by a veritable army of instruments. Here – under a double rainbow – several of the radio telescopes of the 27-strong Very Large Array in Socorro, New Mexico, gang up on the Universe.*

'We were studying quasars, which were new, sexy things. My supervisor Tony Hewish had realised that if we looked for interplanetary scintillation – the rapid twinkling of these objects – we'd be able to pick out more quasars and measure their diameters.'

The 'twinkling' of these distant objects is akin to stars flashing and blinking under the Earth's constantly-shifting atmosphere. The turbulent gas spewed out by the Sun into interplanetary space has the same effect on quasars.

For their research, Bell and Hewish had to build a new radio telescope. In the summer of 1967, a keen bunch of students, including Bell, assembled the structure. It was hardly a conventional dish like Jodrell Bank. Instead, it was a collection of 1000 wooden posts and 120 miles (190 km) of wire, which covered an area as large as nearly 60 tennis courts. Sheep were brought in to graze under the wires, to keep them free of grass.

By November 1967, the 'Four Acre Array' was up and running – and Jocelyn Bell was in charge of operations. 'The output came out on paper charts,' she explains. 'There were very, very few computers – you used grad students instead.'

And if that were not enough, there was the question of having to get to the radio telescopes to make the observations. While the research group was based at the Cavendish Laboratory in central Cambridge, the equipment was 5 miles (8 km) away to the southwest, down a major road that attracts fenland winds by the bucketful.

Bell Burnell is stoical. 'I had a Lambretta scooter. I drove out on this – it was mostly fine, but if there was ice on the roads it was more than a bit hairy. And you got wet, and you got cold, but you got there'.

The young researcher did indeed pick up scintillating radio waves from quasars. But she noticed that there were some altogether different signals. 'Scruff', she nicknamed it.

Many researchers would have dismissed the finding as a technical glitch. But Bell and Hewish were A-class scientists. They rigged up a faster chart recorder, which would reveal just how the radio signal was fluctuating.

On November 28, 1967, Jocelyn Bell watched in astonishment as the celestial signal activated the pen. As if driven by a clock, the pen jumped sideways and back, sideways and back, in a rhythm that was utterly regular – steadier than Bell's own pulse. The period between each 'cosmic heartbeat' and the next was precisely 1.337 seconds.

Tony Hewish's feeling was that the signal was man-made interference. Was anyone in Cambridge operating equipment that pulsed at this unusual rate? No-one was.

Other members of Hewish's team tuned the telescope to a different wavelength and found the pulses coming with the same timing – but a fraction of a second later. No man-made interference would come at different times at different wavelengths. But gases in interplanetary space would delay the longer radio wavelengths in exactly this fashion. The team realised that they had to be looking at a phenomenon way beyond the Solar System.

ABOVE *Twenty-one years after their discovery of pulsars, Jocelyn Bell Burnell and Tony Hewish stand amidst the wiring of the 'Four Acre Array' – the unconventional telescope that located these rapidly-spinning neutron stars.*

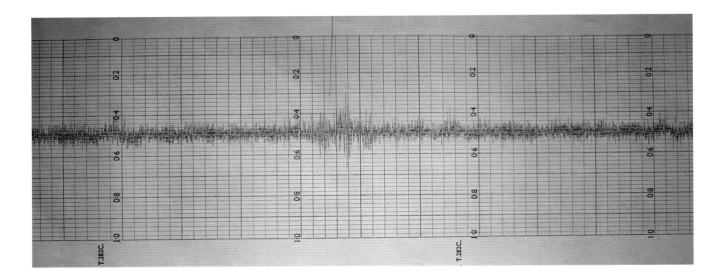

ABOVE *Bell Burnell's 'scruff', seen here on a print-out from the Four Acre Array. The rapid fluctuations in radio signals from objects that the team referred to as 'LGMs' initially baffled the researchers.*

It briefly crossed Hewish's mind that they could be looking at a deliberate signal from an extraterrestrial civilization. In fact, they nicknamed their first source 'LGM-1' – 'Little Green Man-1'. Martin Ryle, the head of the Cambridge department, was genuinely concerned. Ever cautious, he pondered as to whether the extraterrestrials might prove a threat to humanity. And how should they break the news to the world?

Tony Hewish recalls that – if the signals could be proved to come from an alien civilization – Ryle had directed the team to destroy the records at once.

Bell was in two minds about the possibility. She was trying to complete a programme of research into scintillation 'and some silly lot of green men had to choose my aerial to communicate with us!'.

Over the Christmas period, Bell and Hewish worked hard. Bell found three more pieces of 'scruff', in different parts of the sky. They were all beating at different rates. It seemed highly unlikely that four lots of aliens would be trying to signal to Planet Earth at the same time, and in a similar kind of way. Surely these objects had to be some kind of natural phenomenon in the cosmos?

Across in the States, a young Italian researcher, Franco Pacini (who would later become director of the Arcetri Observatory near Florence), had recently published a paper in the prestigious science journal *Nature*. 'It was called *The Energy Emission from Neutron Stars*,' he recalls, 'and in it I suggested that the Crab Nebula could be powered by a rotating magnetised neutron star'.

The Crab Nebula is a torment of tangled gas from the explosion of a massive star that exploded in 1054. By now, it should have dwindled to a pale echo of its former self. But it is still highly energetic, almost a millennium on – and Pacini realised that something must be responsible.

Pacini went to the university library every day to read the newspapers. 'One day, I remember there was a small notice saying that from the cosmos they'd found a signal so-and-so, very precise, etc., etc'. He had picked up on the Cambridge finding.

And he was highly suspicious that the team had discovered the kind of object that he suspected was lurking at the centre of the Crab Nebula.

Jocelyn Bell, understandably, didn't notice Pacini's paper when it came out – it was published at the time when she was actively hunting down the nature of the 'scruff.' She credits him with the recognition that there was a new kind of object in the Universe: one that would solve the Cambridge team's collective bafflement.

'Franco postulated a neutron star in the middle of the Crab Nebula, a rapidly-rotating neutron star with a magnetic field, which was inclined to the spin axis. And that produced magnetic dipole radiation which could not escape – so it was dumped on the nebula, and energised it. He was absolutely spot-on… it was amazing, it really was.'

A neutron star is a cosmic zombie: a star that's died, but still packs a fearsome punch. The collapsed core of a heavyweight star that blows itself apart at the end of

BELOW *The core of the Crab Nebula, as viewed by the Chandra X-ray observatory, is a seething tumult of activity. A pulsar in the centre is whirling around 30 times a second, whisking up the surrounding nebula. The pulsar also shoots out jets of high-speed electrons, to lower left and upper right. Compare this high-energy image of the Crab with the more sedate view through an optical telescope (page 40-41).*

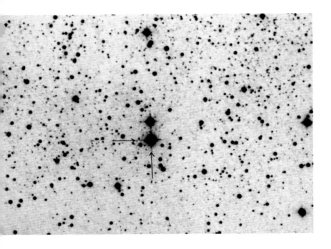

ABOVE *Lair of a black hole: in this negative image of Cygnus X-1 (arrowed), a massive star – 30 times heavier than the Sun – is being relentlessly dragged around by the mighty gravity of an unseen, collapsed object. It is almost certainly a black hole.*

BELOW *An unlikely location for the genesis of black holes: the ancient church at Thornhill in Yorkshire. In the late eighteenth century, the rector – John Michell – theorised that some stars might be so big that light could never escape from them.*

its short life in a supernova explosion, it has no nuclear reactions to hold it up. The atoms of the former star's core meld together into a ball of subatomic particles, called neutrons. You can pack neutrons far more tightly together than atoms – so much so that a pinhead of matter from a neutron star would weigh a million tonnes.

One of the ultimate acts of cosmic violence, a supernova can outshine a whole galaxy – and its legacy lives on in the shape of a ghostly remnant, like the Crab Nebula. But the collapsed core is made of sterner stuff. After Pacini's prediction, researchers discovered the heart of the Crab. And it was beating, like Bell's 'scruff' – at a rate of 30 times a second.

Taking all the evidence together, astronomers realised that they had stumbled over a new species in the cosmic zoo – pulsars. 'We called them 'pulsating radio sources,' recalls Tony Hewish. 'The word 'pulsar' was, I believe, coined by the science correspondent of the *Daily Telegraph*.'

But what kind of beasts were these new kids on the block – objects so powerful that a burst of radiation from a distant pulsar in 1998 damaged Earth's upper atmosphere and disrupted the orbits of satellites?

Pulsars are the most powerful magnets in the Galaxy, each having a magnetic field a million million times stronger than that of our planet. The field traps electrons near its magnetic poles, and the captive particles respond with a cosmic scream for help – sending out powerful beams of radio waves into space. 'The beam is swept around as the pulsar rotates,' adds Jocelyn Bell Burnell. 'It's like a lighthouse. When the beam sweeps across us, we pick up a pulse or a flash.'

A typical spinning neutron star is just 15 miles (25 km) across – no bigger than London or New York – yet it contains more matter than the Sun. This concentrated mass exerts an incredible gravitational pull: 100,000 million times stronger than Earth's gravity. 'Supposing you could land successfully on a neutron star', continues Bell Burnell. 'You'd experience phenomenal gravity. If you tried to climb a little mountain – say, half an inch high – you'd have to do as much work as climbing Mount Everest here on Earth.'

'And the gravity's so strong that the atmosphere is all condensed down to only an inch deep – so it would slosh around your toes. The gravity even bends rays of light, so that you can see over the horizon.'

Now imagine that a supernova left behind a living corpse even more massive than a neutron star. One whose gravity was not just capable of bending light – but of swallowing light whole. The answer: a black hole.

We think of black holes as being at the forefront of all that's new in modern astrophysics. But their discovery – at least in theory – had its origins in a church perched on the side of the valley of the River Calder in Yorkshire. There has been a church on the site since Saxon times, and the present incarnation boasts some of the finest mediaeval stained glass in the county.

In 1783, Cambridge-educated John Michell was rector of Thornhill. The present incumbent is the Reverend Canon Lindsay Dew, a delightful and informal pastor who proudly shows us around his church.

'John Michell was professor of geology at Cambridge, and a maths lecturer, too', he tells us. 'He did lots of experiments in magnetism and geology, and theorized that earthquakes move in waves – which is what we know actually happens.'

'He left Cambridge, because I think he got married. And you couldn't be a professor at Cambridge if you were married – so he became ordained and came up here, because the living at Thornhill was a better paid living than anywhere else in the Diocese.'

Very little is known about Michell. There remain no personal papers, letters, or portraits. The plaque in the church which commemorates him simply reads: 'In the chancel are deposited the remains of the Reverend John Michell, BD, FRS, twenty-six years rector of this parish, eminently distinguished as a philosopher and a scholar. He had a just claim to the character of a real Christian, in the relative and social duties of life, a tender husband, the indulgent parent, the affectionate brother and the sincere friend were prominent features in a character uniformly amiable.'

'I think that's lovely', chuckles Lindsay Dew. 'Uniformly amiable.'

'He spent his time doing experiments down in the Old Rectory. He was a member of the Royal Society, and in November 1783 he gave a lecture to the Fellows theorizing over the existence of black holes. Quite an amazing thing to imagine happening two hundred and odd years ago – to somebody who would have very limited resources and access to scientific instruments. I find it absolutely incredible that someone could have those powers of thought.'

England's Astronomer Royal, Martin Rees, agrees. In admiration, he explains: 'What John Michell worked out was that if you had a body weighing about a hundred

ABOVE *Artist's impression of a black hole in awesome close-up. Glowing matter, pouring in from wrecked stars, ceases to exist as it hits the 'event horizon' – the boundary of the black hole. The hole's powerful gravity distorts the images of stars lying behind it, as predicted by Einstein long before black holes were discovered.*

million times as much as the Sun, light couldn't escape from it. And he went on to say that – for this reason – maybe the most massive objects in the Universe might be invisible to us.'

In 1783, Michell wouldn't have known about the existence of white dwarfs and neutron stars: objects made of superdense material. But as Martin Rees explains, black holes are the natural end point for extremely heavy stars. 'All massive stars die violently, destroying themselves in a brilliant supernova explosion. Most of them leave behind a neutron star or pulsar.'

Rees continues: 'The remains that result from a star that's very heavy – 20 or 30 times as much as the Sun – may instead be a black hole.'

Black holes are the ultimate relics of stardeath. Our dying Sun will eventually bequeath a white dwarf to the Universe; a more massive star will fire up a pulsar; while cosmic superstars will spawn black holes.

A black hole is an object whose material is so densely-packed that even light, travelling at 186,000 miles per second (300,000 km/s), cannot escape its irresistible gravity: so it is black. And it's a hole, because nothing close to the object can avoid falling in, and it can never escape again.

Opinion is continually shifting as to what happens to matter that falls into a black hole. After all – if no light, or other information can emerge – how are we to ascertain what is going on? The odds are that the matter ends up in a pinpoint of infinite density: a singularity. But there are laws of nature that seem to prohibit something that has zero size. And – according to other theorists – black holes could be the gateways to other universes…

A new-born black hole may wander lonely and unseen through the depths of space. But many black holes are not alone. If, in life, the massive star had a companion, then the pair may continue their relationship even after the trauma of the supernova. If the pair are sufficiently close to each other, the black hole can grab matter from its partner – leading to some spectacular effects.

In 1970 – with these thoughts in mind – American scientists launched a new satellite, *Uhuru*, to search for the ultimate fireworks in space. Its job was to track down objects emitting powerful X-rays: energetic radiation that's a sure sign of violent activity in the cosmos.

It discovered hundreds of new X-ray sources. Many of them were neutron stars ripping gas off their companion star. But Cygnus X-1 was different. At the position of this source, in the constellation of the Swan, was a huge, hot blue star some 30 times more massive than the Sun.

It was being dragged around by an unseen object weighing as much as 10 Suns – far too heavy to be a neutron star. The star, and its rogue companion, had already been identified by astronomers at the David Dunlap Observatory in Ontario, but I wanted to see it for myself – sensing that we were on the brink of a high-energy revolution in astronomy.

At the time of the discovery, one of us (Heather) was in a gap year at the Cambridge Observatories before going on to study for a degree in astrophysics at the

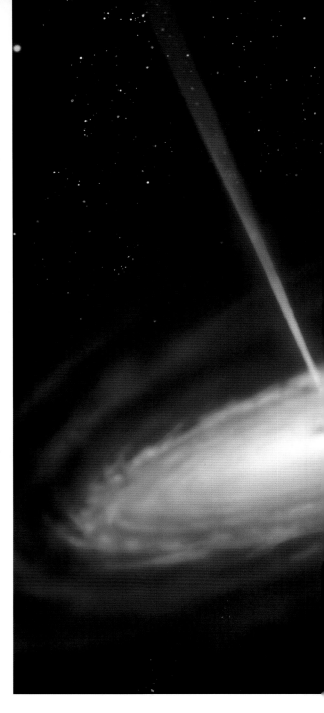

University of Leicester. My inspirations had been twofold. An excellent book on the history of astronomy, *The Universe* – written by Isaac Asimov – and a BBC programme in 1969 called *The Violent Universe*, which was written and presented by Nigel Calder, former editor of *New Scientist* magazine.

It was the first TV programme I had watched in colour and – little did I know it – hundreds of miles away in Belfast, a young Nigel Henbest was glued to the TV screen too. It featured, inter alia, the discovery of pulsars and quasars – plus a great interview with Maarten Schmidt. Unbeknown to one another, we both signed up to become astrophysicists at that point.

On the trail of the mystery star, I remember going into the Cambridge Observatory library with a friend – graduate student Martin Cohen – and poring over the enormous images taken for the complete Palomar Sky Survey. We had the co-ordinates of the source from Uhuru, and searched for the star on the charts. Suddenly, I spotted it.

ABOVE *In this artwork, the mighty gravity of a small but massive black hole drags gas from its blue-giant companion star. The gas swirls around the hole in an accretion disc, where the speed of the captured material approaches the velocity of light. In a last-gasp attempt to escape, some matter close to the hole is spewed out in twin jets.*

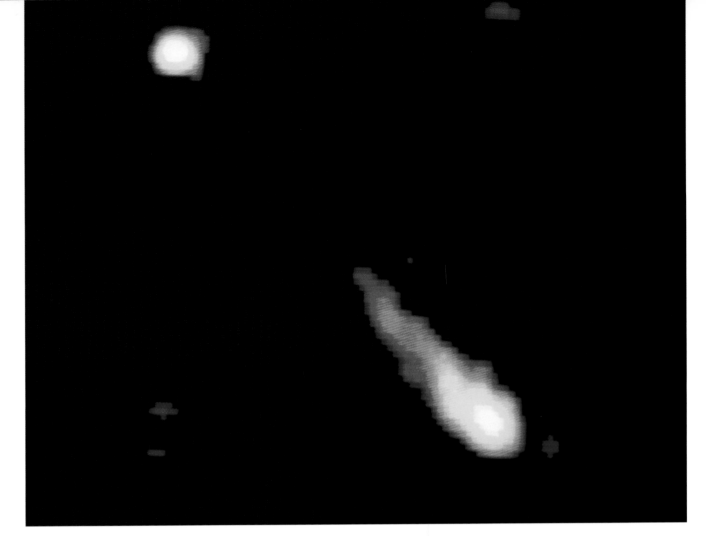

ABOVE *Ultimate black hole: the engine of quasar 3C 273. In this colour-coded radio telescope image, a jet 200,000 light years long rushes away from the accretion disc (top left) of this disturbed galaxy. New research indicates that the mass of its central black hole may be the equal of seven billion suns – amongst the biggest in the Universe.*

Nothing unusual about it. A star imprinted on the photographic plate; so bright it would be just visible through binoculars. But the sheer thrill to think that was being circled by a black hole was awesome.

It was to be the first of many. Orbiting X-ray observatories – way above Earth's blanketing atmosphere – were poised to winkle them out. The satellites homed in on the superheated material that was being dragged off the companion star: the accretion disc. Up close and personal to the black hole, this maelstrom of gases shrieks out powerful X-rays before disappearing from our Universe forever.

And at last, the mystery of the quasars – discovered by Maarten Schmidt in the 1960s, and now with numbers running into thousands – could be answered. 'They're the centres of galaxies which almost certainly contain black holes', observes Martin Rees. 'And they're emitting radiation very intensely because they're being fuelled by gas from their surroundings.'

In essence, a quasar is a mighty accretion disc surrounding a supermassive black hole at the heart of a distant galaxy. And it's no coincidence that quasars are far away and unpredictably violent. They are the hearts of young galaxies – born just after the Big Bang – which are having a youthful feeding frenzy of gargantuan proportions.

But it can't last. Black holes need food. As the galaxy grows up, its core gobbles up the available gas, dust, and torn-up stars in the vicinity, leaving the black hole starved – and its presence unsuspected. That's why there are no quasars in our mature neighbourhood of the Universe today.

Weighing in at hundreds of millions to billions of times the mass of our Sun, the black holes that drive quasars are beasts of a different magnitude from those created by supernova explosions. 'They're monster gravitational bodies that eat their environment,' comments Maarten Schmidt.

A quasar converts the mass of what it is eating into energy. Most of its food disappears down the throat of the black hole, but some manages to escape before being swallowed up. It's the remnants of this wonderful meal that the quasar is enjoying – a giant cosmic burp.

These burps are dangerous. The quasar belches out radiation of all kinds: X-rays, gamma rays, radio waves and brilliant light. The light alone can outshine all the stars in the host galaxy by a hundred, or even a thousand times. And sometimes – as in the case of 3C 273 – the energy creates a jet of electrons thousands of light years long, shooting though space at a speed close to the velocity of light.

Where do these massive black holes come from? Astronomers now think that they're a natural consequence of how galaxies are born. At the centre of a fledgling galaxy, gravity pulls the stars and gas clouds together, eventually creating an enormous black hole.

'It's now believed that almost every galaxy may have a black hole in its nucleus,' confirms Maarten Schmidt. 'Whether you see them or not depends entirely on whether it is eating or not. It may be a billion solar masses, like that in a quasar – or it may be small, like the one in the Milky Way.'

Small is a relative term. Strange goings-on at the centre of our Galaxy – expanding gas rings, bursts of star formation, and a brilliant central radio source – point to the existence of a black hole weighing in at around three million times the mass of our Sun. It's not currently feeding. And at a distance of 25,000 light years from galactic downtown, we're safe from anything that the black hole can throw at us. But if enough straying gas blunders into its thrall, our Milky Way could become a mini-quasar in the future.

Maarten Schmidt recalls the sea-change that took place in our appreciation of the cosmos in the 1960s. 'Until that time you had a very quiet Universe … and suddenly you found explosive or wildly rotating objects like pulsars.'

'You may also remember the film *The Violent Universe* for the BBC – I think it had the right emphasis. The Universe is not as quiescent and calm as it seemed. It was an extraordinary time of change.'

We explain to Maarten that *The Violent Universe* catapulted us both – and a whole generation of youngsters – into astronomy. Like Stanley Hey, he is genuinely bashful and humble. 'Is that really true? Oh, gosh.'

Knowing the producer/presenter Nigel Calder, we get the two men in touch again after a generation of being out of contact. Who knows what a new collaboration might lead to – and how many young astronomers it will inspire to explore our violent Universe?

Are we Alone?

'Let the search commence', declared astronomer Jill Tarter
as she flung the switch on the biggest radio telescope in the world.

ABOVE *The giant Arecibo radio telescope – 1000 feet (305 m) across – nestles
in the Puerto Rico jungle. One of its tasks
is to search for messages from life in space.*

It was Columbus Day – October 12, 1992 – and we were at Arecibo in Puerto Rico to celebrate the official start of humankind's ultimate cosmic quest. The giant dish – 26 football fields in size – had been tuned to listen-in for whispers from intelligent life in the Universe. As the first data began to pour in, there wasn't a dry eye in the house.

'Is anybody out there?' muses Tarter. 'It's the oldest unanswered question our species has posed to itself.'

Modern technology may bring us the answers. But, as Jill Tarter points out, the belief in other lifeforms is nothing new. In the seventeenth century, Dutch astronomer Christiaan Huygens wrote: ' It might nevertheless be reasonably doubted, whether the Senses of the Planetary Inhabitants are much different from ours.' He continued: 'Men reap Pleasures as well as Profit from the Taste in delicious Meats; from the Smell in Flowers and Perfumes; from the sight of beauteous Shapes and Colours.'

Speculating about alien life became popular. Huygens wondered if Jupiter and Saturn were inhabited by great navigators, as each planet has a large number of moons to help guide their ships. And William Herschel – who, in 1781, discovered the planet Uranus – even believed that the Sun was inhabited. In 1835, his son, John Herschel, was the victim of a hoax. While observing previously undiscovered celestial delights in South Africa, the *New York Sun* newspaper maintained that he had actually discovered exotic life on the Moon – leading to a memorable cartoon of naked flying beings.

Some visionaries have even dreamed of making contact with alien life – by signalling our presence to the Universe at large. In the mid-nineteenth century, the Viennese astronomer Joseph von Littrow came up with a scheme that was ambitious, to say the least. He proposed digging trenches in the Sahara Desert in the form of geometrical figures – squares, triangles and circles. Then – at night – these constructions would be filled with a flammable fluid and set on fire, beaming our knowledge of mathematics to the nearby planets.

'It turns out that many brilliant people in the past have thought about contacting extraterrestrials.' observes Frank Drake, the astronomer who is the acknowledged modern father of SETI – the Search for Extraterrestrial Intelligence. 'During my doctoral thesis at Harvard in the fifties, I became intrigued with the possibility that we might really be able to find ET. But back then, the ideas of life in the Universe were not considered very reputable subjects in science. This was all a result of history that had occurred long before: that there were canals on Mars, which turned out to be very bad science.'

Young Frank Drake was not deterred by the orthodoxy. He was at the vanguard of a breakthrough in astrophysics: the development of a new astronomy. Instead of using a conventional optical telescope, Drake surveyed the heavens with a giant radio dish, looking for exploding stars and violent galaxies.

ABOVE *Of all the other worlds in the Solar System, Mars is the most likely to support life. It has volcanoes (left) which belched organic compounds into the thin atmosphere, and ice frozen into its soil. The huge scar across the planet is the Valles Marineris: a canyon system nearly 5 miles (8 km) deep and 2800 miles (4500 km) in length – the width of the United States.*

OPPOSITE *Life on the Moon, depicted in an 1835 cartoon in the* New York Sun. *Despite the size of the aliens' wings, the Moon's airless environment would have a severe problem in supporting these cavorting creatures.*

Then he had his 'aha' moment. Radio waves travel through space virtually unimpeded, as they do on Earth (as you'll know from interference on your sound system!). Could extraterrestrials use a radio telescope 'in reverse' to broadcast signals across the cosmos?

Frank Drake had his wake-up call when he was observing the beautiful Pleiades star cluster. 'I'd done this many nights before. But on this occasion, there suddenly appeared a very strong narrow-band signal, which could only be the product of intelligent activity.'

To check if the signal really was coming from the Pleiades, Drake slewed his dish to a different part of the sky. 'Well – it turned out that when I moved the telescope, the signal was still there. So it was truly from Earth, which was a disappointment. But the seed was planted.'

At the same time that Frank Drake was making his early forays into the world of extraterrestrials, Pete Conrad was training to go there. He was to become an astronaut – and the third human being to walk on the Moon.

We vividly remember the first occasion that a human stepped onto another world. Youngsters at the time, we watched agog at that first lunar landing in July 1969. Neil Armstrong stepped off the ladder with the words: 'It's one small step for (a) man, one giant leap for mankind.'

Four months later, Pete Conrad climbed out of *Apollo 12* onto the barren lunar surface. A man fabled for his sense of humour and infectious grin, Conrad was extremely small for an astronaut – at 5 feet 6.5 inches (1.69 m), he was almost exactly the same height as one of the authors of this book (Heather).

To this day, he's affectionately remembered for his first words on the Moon. 'Whoopie! Man, that may have been a small one for Neil, but that's a long one for me.' Tragically, Conrad died the way that he would have wanted – driving his Harley Davidson along a winding California road at the age of 69, and suffering a fatal crash.

One of the wonderful stories about Conrad is that he underwent survival training amongst the Choco Indians in Panama, and he personally befriended them. After his death, the Choco Chief made him an honorary Choco Indian. In their belief system, the spirits of the dead fly to the Moon.

But his legacy on Earth lives on. *Apollo 12* landed within walking distance of *Surveyor 3* – an unmanned landing craft sent to the Moon in 1967. The astronauts were instructed to dismantle the camera on the craft and bring it home in a sterile container.

Back in the lab, NASA scientists were in for a shock. The camera turned out to be home to a living colony of Earthly bugs, which had probably got into the instrument when a technician assembling it had sneezed.

'I always thought that the most significant thing we ever found on the whole Moon was those little bacteria who came back and lived,' observed Conrad.

How right he was. For over two years, these primitive lifeforms had endured the vacuum and radiation of space, temperatures as low as -420°F (-250°C) degrees, and no access to nutrients or water.

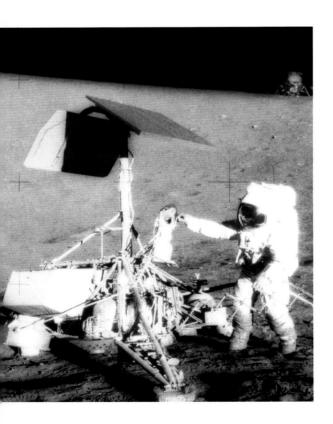

BELOW *Astronaut Pete Conrad, on the* Apollo 12 *lunar mission, inspects the camera on* Surveyor 3 – *an unmanned craft sent to the Moon in 1967. When the crew returned the camera to Earth, scientists were astonished to find that it contained terrestrial bacteria – which had survived the harsh lunar conditions for two years.*

The lesson was clear: life – once established – clings onto it … for dear life.

Life is incredibly hardy. And astrobiologists have recently been discovering just what extremes it can bear. They call the newly-discovered lifeforms 'extremophiles'. These resilient bacteria can live in nuclear reactors, at the bottom of deep bore-holes, in the cold of Antarctica, and in the superheated pools of Earth's hot springs. It seems that life can take in its stride the worst conditions on Earth.

But it needs brave scientists to research and tackle such challenging environments. Enter Jonathan Trent – a biologist at NASA's Ames Institute in California. Using a long pole, he regularly trawls the steaming, boiling pools in Wyoming's Yellowstone Park to collect extremophiles. The tall, bearded scientist warns of the perils of probing into these places out of hell.

'If you fall in, you don't want to be pulled out, because the long and excruciating death from arsenic poisoning would be worse than the quick one that would happen if you're just boiled in the springs.'

So what drives this enthusiastic researcher to take his life in his hands? 'We're going to start exploring for life on other planets,' he explains, 'and the very first thing we should be doing is to understand the extent of living things on this planet.'

Other planets? Which of them are in the frame for the possibility of life? Tiny Mercury, closest to the Sun – and sans atmosphere – is not an option. Venus, second-out, is almost Earth's twin in size. But its thick atmosphere of carbon dioxide has led

ABOVE *In the deep: on the dark ocean floor of the Pacific Ocean, exotic lifeforms thrive around a 'black smoker' – a hydrothermal vent which belches superheated water and minerals into the cold ocean depths, miles below the Earth's surface. Some researchers believe that Jupiter's moon Europa may play host to life like this.*

to a runaway Greenhouse Effect, raising temperatures at its surface to 870°F (465°C) – hotter than any oven setting. Even extremophile enthusiasts rule out these two worlds. But what about the next planet, Mars…?

Mars is fabled for being a living world. In 1877, when Mars hoved close to Earth (as it does every two years), the Italian astronomer Giovanni Schiaparelli noted regular lines crossing the Red Planet's surface. He called them 'canali' – channels.

Word spread to the States, where 'canali' was unfortunately translated into 'canals'. Percival Lowell – a rich Boston banker, and passionate amateur astronomer – took up the baton. The canals, he believed, were the sign of a doomed civilization on a dying planet, desperately committed to bringing water down from the icy Martian polar caps to the desiccated deserts at the equator.

Dedicated to seeking out life on Mars, Lowell built an observatory 7500 feet (2300 m) up in the pine-clad mountains above Flagstaff, Arizona. There, he observed 200 canals – and related: 'They have grown more wonderful with study. They are the most astounding objects to be viewed in the heavens.'

Alas – the canals don't exist. Clyde Tombaugh – the discoverer of Pluto – also used the telescopes at the Lowell Observatory. 'Lowell cut down the telescope's lens with a diaphram to avoid the rainbow colours, so he could get the sharpest images. This makes the bright patches bleed into the dark, making the dark markings a lot narrower. So the canals are totally illusory.'

But belief on Mars dies hard. Lowell wrote extensively on the philosophy of life on the Red Planet: 'They must be globally united and free from the scourge of war.' The

BELOW *Giovanni Schiaparelli's 1877 map of 'canali' on Mars. Percival Lowell, in America, was convinced that these were waterways artificially constructed by Martians desperate to irrigate their dying planet. His opinion led to a widespread belief in intelligent life on the Red Planet.*

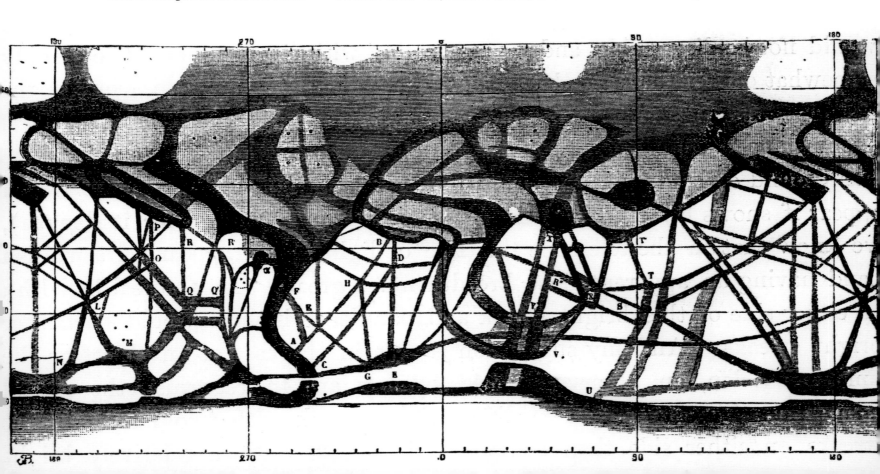

author H.G. Wells, however, disagreed. *The War of the Worlds* relates how Martians flee their doomed world and head for fecund Earth, intent on obliterating all Earthlings and taking the planet over.

His classic sci-fi novel attained immortality on the night of October 30th 1938. Listeners to CBS radio in America were horrified to hear what purported to be a genuine interruption of a regular programme by a continuity announcer. 'Ladies and gentlemen', he intoned solemnly, 'I have a grave announcement to make. The strange object that fell at Grovers Mill, New Jersey earlier this evening was not a meteorite. Incredible as it seems, it contained strange beings who are believed to be a vanguard of an army from the planet Mars.'

The 'continuity announcer' was none other than the 23-year-old actor and director (and sometime bullfighter and magician) Orson Welles. He had a responsibility to put on plays on Sunday nights to boost the flagging ratings of CBS. His ruse certainly worked this time – and drove terror into the heart of the American nation.

A voice from Washington revealed that Martians were landing all over the United States. At one point, Welles announced that 'a reporter in the field' described a Martian emerging from its spacecraft. 'There, I can see the thing's body – it glistens like wet leather. But that face. It… it's indescribable. The mouth is V-shaped, with saliva dripping down from its rimless lips…'

Washington came back with a report that thousands of people had already been massacred by the Martians with the aliens' death-ray guns. Their fighting machines – the Tripods – were everywhere. And eventually, they were to converge on the CBS studios in Manhattan – where the broadcast ended abruptly in a chilling, high-pitched scream.

The acting was so convincing that many Americans had no idea that they were listening to a dramatisation of H.G. Wells's novel. They fled their homes, hid in cellars, prayed, and even wrapped wet towels around their heads to protect themselves from the noxious Martian gas.

Welles was blissfully unaware of the havoc he had wreaked until he bought a newspaper the following morning. 'Radio listeners in panic' screamed the headline. 'Many flee homes to escape gas raid from Mars.'

Orson Welles' career never looked back. And the concept of life on Mars was now firmly on the agenda.

So it was hugely disappointing that the first spaceprobes aimed at the Red Planet in the 1960s discovered a bleak, moonlike world. Mars was barren, peppered with craters, and its thin carbon dioxide atmosphere had a pressure less than one-hundredth of that of the Earth. The planet appeared to be completely dead.

But later probes started making researchers suspicious. There were tantalising signs of recent volcanic activity. And newly-discovered dry, winding channels bespoke of past water on Mars. Perhaps the planet was once a living world. Might it be alive today?

Seven years to the day after Neil Armstrong and Buzz Aldrin had landed on the Moon – July 20, 1976 – the most sophisticated robot probe ever designed touched

ABOVE *A contemporary illustration of a Tripod – a Martian fighting machine – from H.G. Wells's 1898 novel* The War of the Worlds. *Orson Welles's later radio dramatisation of the book in America fuelled panic through the nation, when listeners really believed that Martians had invaded New York.*

down on the Red Planet. *Viking 1* was joined on the surface of Mars by its sister-craft, *Viking 2*, on September 3.

The geologists and biologists in the astronomical community had their own, separate, agendas for the *Vikings*. Were the probes going to look at the Red Planet's unique landforms – or were they going to search for life? In the end, the biologists – led by the iconic Carl Sagan – won. This meant that the mission came in with a whopping price-tag of $2 billion. The probes had to be expensively heat-sterilized before launch to remove earthly bacteria, and to avoid contaminating any Martian life.

Back on Earth, we were treated to stunning images of an alien world – a planet with rusty-red soil strewn by dark rocks, under a canopy of a salmon-pink sky. Both landers returned weekly weather reports, analyses of the Martian air, and wind-speed measurements.

But not everyone was as sanguine as Sagan that the *Vikings* would detect life on Mars. One notable detractor was the independent, highly-respected British scientist, Jim Lovelock. Famed for his Gaia hypothesis – that our planet has evolved to shape living beings, and that life, in turn, has shaped the Earth – Lovelock is no touchy-feely environmentalist.

'I'm a cussed person, and I wanted to go independent,' he says. 'I work from my own laboratory, and I fund myself by my wits.'

Totally charming and mischievous, Lovelock is the epitome of what it is to be English. Yet the American space agency NASA picked him over 40 years ago to become one of their senior advisors. On *Viking*, he was an instrument scientist – and

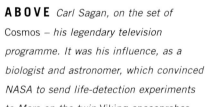

ABOVE *Carl Sagan, on the set of* Cosmos *– his legendary television programme. It was his influence, as a biologist and astronomer, which convinced NASA to send life-detection experiments to Mars on the twin* Viking *spaceprobes.*

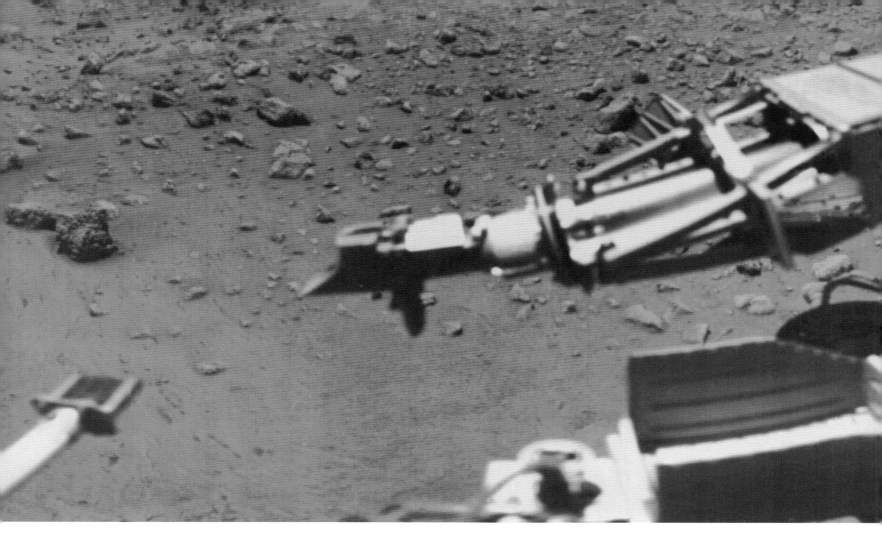

deeply sceptical about the possibility of life on Mars. 'If you go back to my *Nature* paper of 1965 – a hell of a long time ago – I said that the best way to look for life on Mars was to analyse the atmosphere. But all the life experiments they were sending to Mars were so geocentric that I didn't think they had they'd have a dog's chance of working. And so I started criticising, and suggested they should look at the whole planet – instead of just something on the surface.'

But the biologists had the last word. They designed a miniature Martian laboratory – the size of a wastepaper bin – with a ten-foot-long arm to feed soil samples into the bellies of the *Viking* landers.

There were four main experiments. Three were designed to look for signs of life in the soil, while the fourth – the Gas Chromatograph Mass Spectrometer (GCMS) – broke down the soil into its basic atoms.

The GCMS detected no carbon atoms in the Martian soil; and if there's no carbon, there's no possibility of life. So NASA's official line is that there is no life on Mars.

Scientists came to the conclusion that Mars had been rendered sterile by ultraviolet radiation from the Sun. The planet's almost airless environment means that it has been the victim of our local star's worst vicissitudes over billions of years. Add to this the fact that Mars's atmosphere contains virtually no oxygen – so there's no protective ozone layer – and life has no hiding place.

But we became intrigued by one of the experiments that seemed to contradict the orthodoxy – which is how we came to meet up with a sanitary engineer at a light industrial estate in Beltsville, Maryland.

ABOVE *Self-portrait of the* Viking 1 *lander, exploring Chryse Planitia on a summer's day in August 1976. The scoop (right) was used to dig trenches in the Martian sands to a depth of 12 inches (30 cm), and test the soil for signs of life.*

OPPOSITE *The respected independent scientist Jim Lovelock is anti-life on Mars – he believes that the composition of its atmosphere proves that there is no life on the Red Planet. Here he relaxes at the gate of his mill house in Cornwall: the Coombe Mill Experimental Station.*

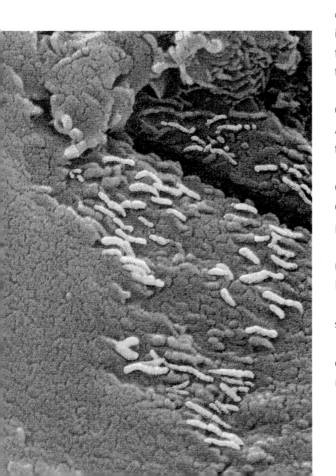

Gil Levin – CEO of Biospherics Inc. – was the member of the *Viking* team whose package on board hinted at life. 'I developed a method to detect micro-organisms very quickly, as a result of a need I saw when I was a public health engineer,' he explains. 'I was an investigator on the Labelled Release experiment. That's the one that got a positive indication for life on Mars, and has kept me in trouble ever since.'

It's fair to say that most of Levin's colleagues are extremely sceptical of his findings – partly because Levin is not a trained biologist. But now, some are rallying round.

Levin describes his technique, based on the kind of devices he's built to track down Legionnaire's Disease in air-conditioning systems. 'It's very simple. The standard method of culturing micro-organisms is to put them in some kind of nutrient soup, and to wait several days until they start multiplying. My technique simply added radioisotopes to those nutrient compounds. That meant that as soon as the micro-organisms started metabolising them, they would expire radioactive gas.'

And Levin's experiment yielded up copious quantities of gas. One researcher in the control room reported 'My God, my God – there may be life there. And then of course, it all kind of waned.'

Actually, this wasn't true. Levin showed us his data, which show the release building for two days, and then continuing for the full eight days of the programme. And the graphs looked identical to his measurements of biological activity from Earth-based bugs.

Levin went further. He heated the soil on Mars to a range of temperatures that would kill off bacteria on the Earth. 'First of all, we showed that 51 degrees definitely destroyed the signal. But we showed that 46 degrees didn't destroy it – it inhibited it by 30 per cent. And that's just the way that in the lab here we distinguish E. coli from the rest of the coliforms, because E. coli can survive beyond 37 degrees, while the others cannot.'

But NASA had made its collective mind up that there was no life on Mars. The GCMS had revealed no carbon, and Gil Levin's outspoken claims turned him into an Agency exile. 'It was political,' he tells us. 'They had to come down with a decision, and they hate to retract a decision.'

On the opposite coast of the States, we visited the Scripps Institution of Oceanography at La Jolla in California. A team there is assembling a new version of the GCMS to break down the Martian soil and analyse its atoms. They seemed genuinely perplexed at NASA's verdict that there was no carbon on Mars.

'You have to remember that the *Vikings* were in 1976,' points out one of the researchers. 'We've recently done some really interesting experiments with our new Mars Organic Detector. Comparing the GCMS with our instrument, we estimate that *Viking* would have missed out on the order of 30 million bacteria cells per gramme of soil. So there could have been cells in the soil, but *Viking* wouldn't have seen them.'

And there have been more tantalising hints that life on Mars has existed, and still exists today. In 1996, a team from the Johnson Space Center in Houston analysed a meteorite that had fallen on the pristine snows of Antarctica. From its composition, scientists knew that the rock had been blasted out of Mars by a collision with an asteroid – later to land on the Earth.

The researchers took ALH 84001 to pieces. And they were astonished to find that it contained what appeared to be microscopic, fossilised 'maggots' inside the rock. The finding provoked a furore. Newspaper headlines screamed out 'Life on Mars – official,' while President Bill Clinton proclaimed 'Today, rock 84001 speaks to us across all those billions of years and millions of miles. If this discovery is confirmed, it will surely be one of the most stunning insights into our Universe that science has ever uncovered.'

The plethora of probes circling Mars, and roving its surface, have also turned up more circumstantial evidence for life. The orbiters have located gullies where water may have flowed recently. The rovers *Spirit* and *Opportunity* have also homed-in on sites whose geology bespeaks of liquid surface water in the past. And Europe's orbiting Mars Express probe has picked up traces of the gas methane emanating from the Red Planet. This could be result of vulcanism – or it could be a result of life.

It has to be said that the jury is still out when it comes to the question of life on Mars. But could there be life farther out in the Solar System? Spacecraft probing the giant outer planets have come up close and personal to the enormous moon-systems circling Jupiter and Saturn. Some astronomers believe that they are more likely to be the abode of life than the Red Planet itself.

Centre-stage is Europa, which – at 1900 miles (3000 km) across – is one of the largest moons orbiting Jupiter. It has an incredibly smooth surface, which astronomers interpret as a layer of ice. Because Europa is being gravitationally pummelled by Jupiter, it must be warm inside – which leads researchers to believe that the moon has a deep ocean under the icy crust.

'I think the ocean's a fascinating place,' observes Frank Drake. 'It has the potential for having more complicated life forms than on Mars. I think all the life forms on Mars will be extremely primitive single-celled organisms, but on Europa there's been an opportunity for evolution to occur in the ocean. If there's life there, you might find some very interesting creatures.'

Only 300 miles (500 km) in diameter, Enceladus is the baby sister of Europa. One of Saturn's mighty clutch of moons, this billiard-ball-smoothy of a world is made entirely of ice – and yet has erupting geysers. The combination of heat and water make for an ideal habitat for life.

And then there's Titan. The largest of Saturn's moons (over 3000 miles / 5000 km across), it is enveloped in an orange atmosphere of nitrogen nearly twice as dense of that of the Earth. In January 2005, the European *Huygens* probe – named after the Dutch discoverer of the moon – landed on its surface. It found a world quite unlike anything else in the Solar System. It appears to have lakes of liquid methane or ethane, plus hot spots of volcanic activity.

Taken together, these are the basic conditions to create life. But – at its distance from the Sun – Titan is too cold to support living beings. Yet in five billion years time,

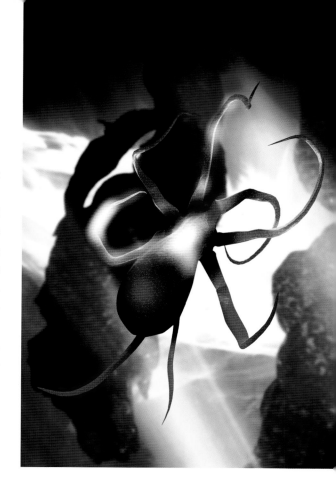

ABOVE *Arthur C. Clarke's novel* 2010 *first explored the possibility of life on Jupiter's moon Europa. Under its girdling ice-sheets, there appears to be a warm ocean. Could a creature like this be one of the denizens of the deep?*

PREVIOUS PAGE *Over 200 planets circling other stars have now been ferreted out. The artwork depicts a gas giant, like Saturn, in the fledgling Orion Nebula – as seen from one of its moons.*

BELOW *The dense atmosphere which envelops Titan - Saturn's largest moon – conceals a surface which boasts lakes of liquid methane, in which life could develop.*

when our local star billows out to become a red giant, primitive life on Titan could have its day – before petering out as the Sun dwindles.

But our Solar System, despite its wonderful diversity of worlds, is only the tiniest corner of the Cosmos. And the potential for life 'out there' appears to be vast. One of the most incredible revolutions in astronomy has taken place in the last two decades: the discovery of planets around other stars. Over 200 have now been found.

No-one has actually seen an extrasolar planet directly. The problem is that stars are big and bright; planets are small and dark. It's as if you're trying to detect moths around a streetlight when you're stationed in London, but the streetlight is in New York. The world's leading planet-finder, Geoff Marcy, explains the secret to his success: 'The technique is to watch the star carefully. Then see if it wobbles due to the gravitational pull of its planets.'

Marcy recalls what drew him to searching for planets around other stars. 'I was 28, and doing research into the magnetic fields of stars. Frankly, it was not all that grippingly exciting. My career was not even exciting to me – so how could I expect it to be exciting to my colleagues or anyone else?'

Taking a shower in his Pasadena apartment a few days later, Marcy realised the way forward. 'I needed a project that gripped me on a very personal, almost childlike level. I thought to myself that I should try to find a project that I had asked myself as a child. And the question that popped into my mind, with the shower still running, was whether or not there were planets around other stars. That's how it all got started.'

As it happened, Marcy was pipped to the post by Michel Mayor and Didier Queloz from the Geneva Observatory, who'd sussed out the same 'wobble' technique. On October 6th, 1995, the Swiss team announced that they had discovered a planet orbiting the Sun-like star 51 Pegasi.

To put it mildly, young Didier Queloz was amazed. He was expecting a very slow change in the speed of the 'star wobbles.' Even with their new precision measurements, they knew that they could only pinpoint planets roughly as massive as Jupiter and Saturn. And – in our Solar System – these giant worlds lie a considerable distance from the Sun, so that the 'wobbles' they create take years to complete.

'I saw something having this variation of four days,' Queloz recalls of 51 Pegasi. 'I had no other explanation than there was something orbiting it. If I had been older, I would have said – oh no – it cannot be a planet.' But Queloz continued his observations, and came up with the inescapable explanation: 51 Pegasi was being circled by a planet half the mass of Jupiter, orbiting in just 4.2 days – so close to the star's surface that the pair must almost be touching.

Marcy, who had been on the trail of extrasolar planets for eight years by then, was staggered. Paul Butler, Marcy's colleague, was equally stunned. 'We thought – four days! This must be crazy – we've never heard of such a thing.'

But after an observing run of four nights, Marcy and Butler were able to confirm the Swiss team's data. 'There was so much skepticism about this claim,' remembers Marcy.

'We thought we should just make it publicly available. So we took our plot of data, and simply put it onto the Internet – we just made it available to the whole world.'

This was the wake-up call for Marcy and Butler. Over the previous eight years, they had been looking for long periods of wobbles in their target stars, assuming that the planets responsible lay in remote, slow orbits.

So over the next two weeks, they re-analysed their data – and were rewarded with the discovery of their first two planets: one circling the star 70 Virginis, and the other in orbit around 47 Ursae Majoris. Both were in relatively short-period orbits, circling their parent stars in 116 and 1089 days, respectively. Another member of Marcy's team – Debra Fischer – discovered another, more distant, planet circling 47 Ursae Majoris in 2002.

By now, the planet hunters had the bit between their teeth. Astronomers had for so long been on the quest, but the technology wasn't good enough. Now it was up to the task – and planets started to tumble into the telescopes like snow in a blizzard. But the discoveries raised more questions than ever expected...

For a start, the wobble technique can only detect worlds which are roughly as massive as Jupiter or Saturn – in other words, gas giants where life would find it impossible to gain a foothold. And as Geoff Marcy points out, these planets don't obey the orderly structure of our own cosmic neighbourhood. 'The majority of them reside in wacky oval orbits. There are some close in, some far out.'

And the discovery of massive planets up close and personal to their parent star is particularly disturbing. These 'hot Jupiters', with their atmospheres boiling away under the searing heat, are killers when it comes to the possibility of life in other solar systems. Wreathed in massive atmospheres, these worlds must have been born far from their sun. Only in a cool part of the embryonic nebula surrounding the young star could they have managed to snatch the gas that now envelops them.

ABOVE *Michel Mayor (left) and Didier Queloz (right) celebrate their discovery of the first extrasolar planet by reading their paper in the renowned journal* Nature.

BELOW *A thoughtful Geoff Marcy ruminates on techniques to discover planets around other stars. To date, he and his team have been the most prolific planet-hunters.*

ABOVE *Leading SETI astronomer Jill Tarter stands in the focus box of the Arecibo radio telescope, 450 feet (140 m) above the huge dish – the biggest radio telescope in the world. It is one of many vast telescopes that are hunting for extraterrestrial signals.*

Then, for whatever reason, they careered inwards. In doing so, they would have driven any smaller, rocky terrestrial planets like the Earth – the worlds capable of bearing life – into the parent star.

The new findings are making astronomers revisit all their ideas on planetary birth. 'It seems our Solar System is some kind of special case,' observes Marcy. 'It has just the orbits it needs to have – i.e. circular – for life to proliferate.'

But some recent discoveries have found 'Jupiters' in stable orbits at just the right distance from their parent star. Could there be small, rocky planets with life on their surfaces closer in?

Alas – our present technology doesn't allow us to winkle out Earth-sized planets, but Geoff Marcy doesn't think it will be long before we'll do it. 'In 15 or 20 years, we'll have space-borne interferometers. The idea behind them is brilliant. If you combine the light waves from a star – those that interfere and cancel out – you can get rid of the starlight. So a planet like the Earth could be made visible against this nulled-out starlight.'

'What's extraordinarily promising is that the interferometer could take a spectrum of light from a planet – and we could analyse it to search for the constituents of that planet's atmosphere. Probably the most telltale and amusing of them is methane gas, which is produced – frankly – by bovines out of their backsides.'

The discovery of extrasolar planets was a real spur to the SETI astronomers – although there were trials and tribulations along the way. The community's NASA-funded search at Arecibo in 1992 was dead just a year later. It was killed off by Nevada senator Richard Bryan, who derided the $100 million project – which amounted to only a fraction of a single Space Shuttle launch – as 'a great Martian chase'. Unbelievably, the US Congress supported him. The *Boston Globe* newspaper criticised the decision spot on: 'It proves one thing, and one thing only. That there is no intelligent life in Washington.'

Jill Tarter despairs at the misconceptions that surround SETI. 'There are so many wonderful things about the job I get to do, but there are frustrating things as well. One is the lack of critical thinking skills among the general population, particularly in the United States. We have worked very hard to set up a scientific research programme that will be credible, verifiable and repeatable. And to have people not appreciate the difference between that and someone who says 'You know, I saw an alien in my bedroom last night. 'Course you can't see him, but I can.'

However, the SETI researchers were made of stern stuff. They organised themselves into business entrepreneurs, created a private company, and set about an international fundraising campaign. After 15 months, the team was several million dollars richer – and Project Phoenix was born, controlled from the newly-founded SETI Institute in Silicon Valley, with Jill Tarter (famously portrayed by Jodie Foster in the movie *Contact*) at the helm.

The group began by taking their detection equipment to Australia, to be bolted on to the giant Parkes radio telescope in New South Wales. And – twice a year – the Phoenix team returns to Arecibo, trying to eavesdrop on aliens hundreds of light years away. A mere 5000 miles (8,000 km) across the Atlantic, the giant Lovell Telescope at Jodrell Bank also gets a slice of the action. The two dishes work simultaneously to confirm – or reject – each other's detection of a suspected signal.

And now they have the beginnings of their own radio telescope facility. At Hat Creek in Northern California, an array of 20-foot (6-m) dishes is springing up, to listen in to the call from space. It's hoped that the final telescope will grow to 350 dishes.

But after nearly 50 years, the team still hasn't received the phone call from ET. At the SETI Institute, however – on a beautiful day at the height of California's summer – founder Frank Drake doesn't look too despondent.

It's early days yet, he feels. Maybe looking for radio signals is a primitive way to track down extraterrestrials. Might they be using laser beams instead? Or is ET a strong and silent type? After all, humankind has done little to advertise its presence to the Universe at large.

Drake recalls some of our first attempts. The first was at Arecibo, in 1974, when he was director of the giant radio telescope. The dish had been resurfaced, and Drake wanted a ceremony to celebrate the event. His PA came up with the brilliant suggestion: 'Why don't you use it in reverse and beam a signal to the Universe?'

'In the middle of the celebration,' Drake recalls, 'we transmitted to the stars in the direction of the great globular cluster M13. This is a group of 300,000 stars 25,000 light

BELOW *A phoenix arises from the ashes: the SETI team, having had their original search killed off by the US Congress, re-launched it as 'Project Phoenix' in 1995. Their first observing run was conducted on the 210-foot (64-m) diameter Parkes telescope in New South Wales.*

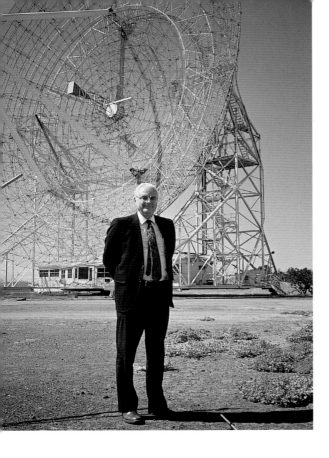

ABOVE *'Father of SETI', Frank Drake, with an early radio telescope in Silicon Valley. Drake is optimistic that there are many intelligent civilizations in the Universe, and that some of them might be immortal.*

BELOW *Message to another lifeform: the plaques on the* Pioneer 10 *and* 11 *spaceprobes, which are leaving the Solar System, reveal the geography of our location in space, and the appearance of the creatures that sent them.*

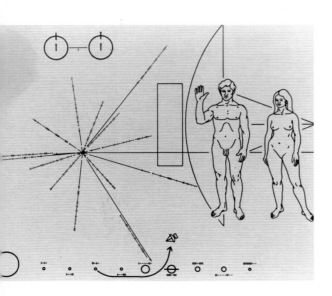

years away from us. The message was made in a code that was easy to break. We showed the basic chemistry of life on Earth, the DNA molecule, and the arrangement of our Solar System. And we gave our population, Earth's size, and the size of the telescope which sent it.'

Drake, in collaboration with Carl Sagan, went further. They designed a plaque on the sides of the two *Pioneer* spacecraft destined for Jupiter and Saturn, and which are now heading out of the Solar System. 'They have crude line drawings of human beings, and they locate the Earth with respect to 14 nearby pulsars.'

But the design of the plaques didn't go down well with everyone. 'The two human figures on the plaques are both nude,' explains Drake. 'We thought: the extraterrestrials might be interested in just what our anatomy is, which is why humans are that way'. The citizens of middle America were not amused. 'In the *Los Angeles Times* there were letters to the editor protesting that we were using taxpayers' funds to send smut into space.'

Drake had to be more circumspect on his next attempt at communication. The twin *Voyager* craft – also aimed at the outer planets, but destined to escape towards the stars – contained an old-fashioned LP record on their sides, complete with a stylus! The discs were encoded with sounds, pictures and messages from Earth.

But in selecting them, Drake and his team came up against a very sensitive NASA bureaucracy. 'It was obvious that we'd show nude humans – so we had to use a picture from a medical handbook. We're actually embarrassed that the extraterrestrials will giggle – because they'll recognize what has happened. And they'll recognize that in our civilization, such prohibitions, such hang-ups, still exist.'

Robot messengers like the *Pioneers* and the *Voyagers* may be the way to carry our thoughts, our senses and our knowledge into the community of extraterrestrial life – but they're no substitute for going there ourselves.

In fact, the way to the stars – and, hopefully, the discovery of life – may not come from looking to the future, but by looking to the past. The Polynesians have a lot to teach us, according to anthropologist Ben Finney. Their epic voyages across the Pacific Ocean were driven by a passion to find the next island: stepping-stones in their quest for discovery.

Finney sees a time ahead when our descendents – who will have been progressively colonizing the Solar System – will reach a point of no return. 'They will undergo a change in consciousness analogous to that experienced by the seafarers who colonized the Pacific Islands, but one of infinitely greater import for the future of humanity.'

Ed Krupp, of the Griffith Observatory, Los Angeles, is absolutely in sympathy with Finney's vision. 'We live in an era where we have transformed the sky. We don't just look into it; we go into it and we bring it back to us.'

'And this is a time when we disperse the sky through our media with discoveries of astronomical consequence that are every bit as mystifying, every bit as imaginative, every bit as compelling as the sky was to our ancestors.'

As an historian of ancient astronomies, Krupp has a vision of the past millennia of astronomical discoveries that he projects forward to the present – and to future millennia.

'We get a three-dimensionality of the Universe that I didn't even have as a kid. We've left the surface of Earth, landed robot probes on the surfaces of other worlds, put telescopes in orbit around our planet, we've gone to the Moon and brought back rocks, turned planets into landscapes, and now we're turning stars into solar systems.'

'And these vistas that were once black-and-white framed pictures of stars and nebulae are now these extraordinary sensibilities that – when you look at them – you just want to immerse yourself in a picture of incredible depth. And we find ourselves actually on the springboard of the entire Universe.

Astronomers have always been at the cutting edge of our greatest intellectual advances – from the realisation that our so solid Earth is actually careering through space, to the mind-numbing properties of a black hole.

And this voyage of celestial discovery is far from coming to an end. In this generation and the next, we are experiencing a new revolution in the history of astronomy – one that directly connects humankind with the Cosmos.

BELOW *Just as our Polynesian ancestors used the stars to navigate the Pacific to destinations such as Easter Island, so might we – in the future – rely on our knowledge of the heavens to take us to the farthest reaches of the cosmos.*

Index

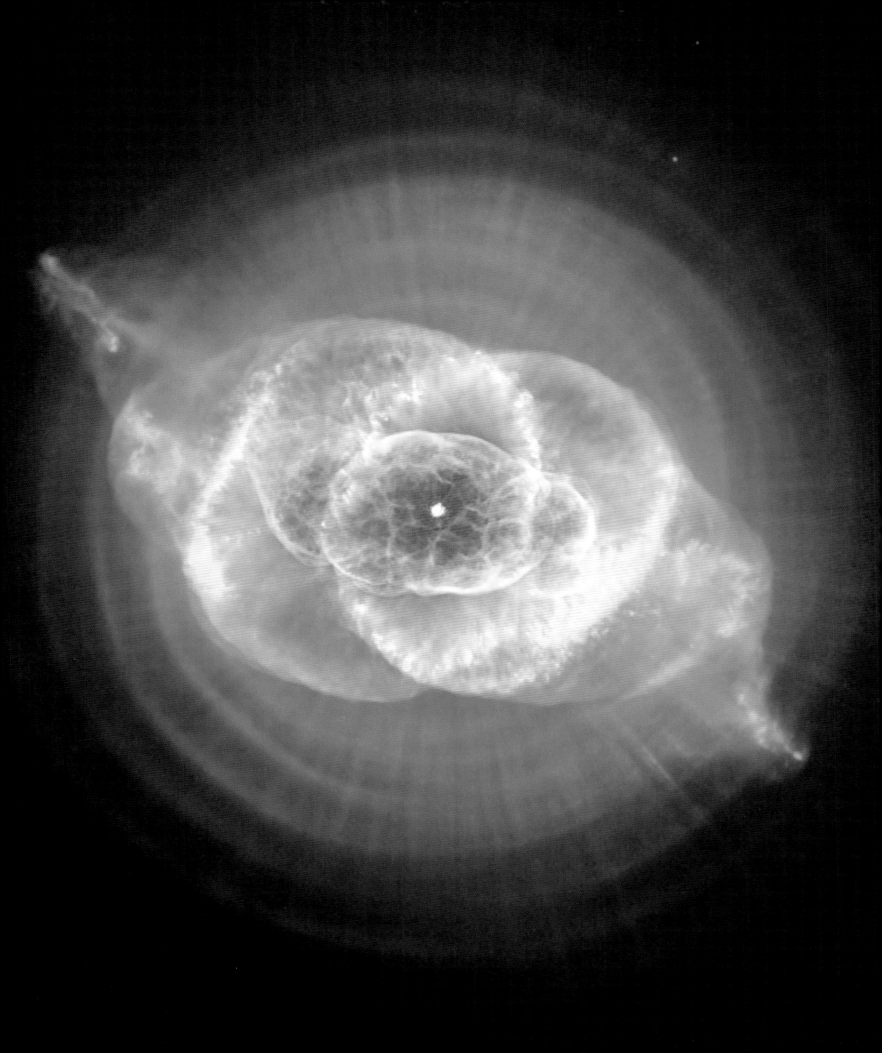